The Rhetorical Resource
Irony

Jonas Batista dos Santos

Copyright © 2016 - Jonas Batista dos Santos

All rights reserved.

ISBN: 9798241464347

Issue: 10

Review: 20260513

DEDICATION

I dedicate and thank God for everything, especially for having given me my children Maira and Pedro.
I am also grateful for the people who love me.
Lord, thank you for your grace.

Laudate Dominum.

CONTENTS

I - JUSTIFYING PROPOSITION 1
II - ANCIENT GREEK IRONY 11
    The Playful and the Misleading 39
III - THE BREADTH OF THE DEFINITION 55
IV – CRITICAL ANALYSIS 77
    Characteristics of the Literary Genres Short Story and Novel 77
    Historical Contextualization - The Illuminist Voltaire 83
    Candide or Optimism 95
    Memnon or Human Wisdom 131
    History of Escarmentado's Travels Written by Himself 143
V - CONCLUSIONS 157

I - JUSTIFYING PROPOSITION

I propose to discuss, through this writing, the most complex and least understood rhetorical resource, the Greek irony rhetorical resource of the Ancient Age with the root of its excesses added to its definition whose importance I consider for the understanding of the phenomenon, which can bring a firm basis to provide assertiveness in understanding and using the resource.

The universal definition combined with the definition of its referent in the form of a written manifestation provides the necessary sample for understanding the complete manifestation of the phenomenon, so exposing a sample of the Greek phenomenon present in the literature of the Ancient Age is an essential action when discussing the phenomenon.

A comparison between the manifestation of the rhetorical resource Greek irony of the Ancient Age, its manifestation in the final period of the Modern Age and its manifestation in the Contemporary Age, in literary productions and examples that allow the study of the breadth and excesses of the definition of the phenomenon, should provide what is necessary to form the idea of the original resource and distinguish it from the excesses that can be predicated in many ways, in addition to providing what is necessary to train, detect and produce the resource in literary works, as well as assist in defenses and detection of mistakes, not having as a guarantee the correct human judgment about who wins when using the resource.

As a sample of ironic ancient Greek text, Plato's *Euthydemus*, written in ~384 BC, is used in this writing, which makes use of irony and sarcasm. In it there is a sample of the ancient Greek phenomenon in the search for truth, virtue and wisdom, as well as in the search for success, false reason or ridicule to promote mockery, mockery and derision in certain cases adequate or not to the truth.

Adequacy must be made between the phenomenon, its semantic load and its referent, and there is a sufficient sample of the phenomenon for this in the proposed dialogue.

Obviously, many ancient Greek dialogues have samples of the phenomenon for analysis, however, in the proposed one, there is a clear distinction in use depending on the individual character where the individual who knows his own limitations and seeks the truth develops the phenomenon in a very specific way, as he distinguishes himself from those who seek only playfulness or deception for the sake of personal success. In Euthyphro, for example, we would not have a sample of opposition to be able to compare ironies depending on character, which we find in *Euthydemus* in a complete way.

Part of the semantic load of the phenomenon in contemporary times may be the basis, even if mistaken, for the analysis of the ancient phenomenon, but an analysis must conclude why the definition of 'irony' has undergone many additions and why it is necessary to develop criteria when analyzing it with the use of its additions.

In order to highlight and analyze the phenomenon present at the end of the Modern Age, because the Modern Age was the period after the Middle Ages that rescued Greek culture through Catholic universities and also because the final period of the Modern Age included medieval and modern culture, this paper proposes to study the rhetorical resource irony in the novel published in 1759, at the end of the Modern Age, *Candide or Voltaire*'s Optimism, which is very wealthy when it comes to ancient Greek irony. The short stories *Memnon or the Human Wisdom* and *History of the Travels of Escarmentado Written by himself,* by the same author, will also be analyzed in a more succinct way , as they have similar proposals and points of convergence that allow the observation of common elements in different forms, which leads to evidence of the studied resource manifested in its literary forms in the literary genres novel and short story.

The detection of the rhetorical resource irony depends on the correct interpretation of the narrated phenomena, which requires pertinent historical contextualization that provides data on the context of the analyzed works and, in order to understand what is narrated, it is necessary to understand the values linked to what is narrated through history, which generates the need to briefly discuss the education of the period that comprises the Middle Ages to the end of the Modern Age, the education received by Voltaire, to contextualize the reading and understand aspects of Enlightenment thought and Voltaire's thought, which allows us to understand, by making use of assertive reading, part of the ironies in the proposed corpus with the evidence of some elements of the period that help to understand the complexity of the phenomenon and the genius of the great author of French literature.

A part of the books on Voltaire and the reproductions of his works have rich information on the dating of serialized events in his life that serve to form some conclusions and create a link between the man Voltaire and his productions, which is necessary to understand the origin of his great writing skills, a taste for literature, personal inclinations and is essential in an analysis that perceives his great novel *Candide or Optimism* as a continuity of the way the author acted in his current and real life.

This makes it possible to have indications of the reality, of its semantic modifications and situational contradictions, which can allow us to know whether they are the result of literary art for the playful or the result of erroneous personal convictions or not, to evidence a supposed truth, among other possibilities.

The proposed works have the rhetorical resource irony in a very complete way, so they allow the in-depth study of the phenomenon by the original Greek definition of 'ειρωνεία' (irony) in comparison with its breadth and additions acquired over time, in addition, they have similar proposals, connections and shared elements that will be evidenced in critical analyses.

To exist the notion of the expanded or exceeded definition, quotations from the theorist D. C. Muecke, who was a scholar of the phenomenon and developer of the work *Irony and the Ironic*, are also used. Through this theorist, I bring the definition of **verbal irony** or **instrumental irony** and that of **situational irony**, to allow the development of an understanding of the manifestation of the phenomenon studied in the literature in order to highlight the reason for the additions to the definition by D. C. Muecke having theorized about the phenomenon during his writing in the year of his death, 2015. The **dramatic irony** will also be evidenced when there are timely manifestations that contribute to the proposal.

It is possible to say that the novel *Candide or Optimism* proposes to be ironic and Voltaire, who was privileged to have studied through the Catholic Church that rescued classical Greek education, makes sufficient use of the Greek resource and its amplitude, if amplitude is considered as evidence of the phenomenon, which makes the novel and short stories an excellent corpus for the analysis of the rhetorical resource irony.

In his great novel *Candide or Optimism*, Voltaire playfully mocked the theory of the best of all worlds, by the connoisseur of many sciences Gottfried Wilhelm Leibniz, a theory that sheds light on any phenomenon in the world due to its universality.

Voltaire kept the focus on the great questions of his time, which include the conceptions of good, evil, morals, ethics, religion that includes Christianity, Protestant sects and others, philosophy and, within this, metaphysics, but he could simplify and say that the targets of his ironies are only philosophy and religion, because they are great sources of guidance and many govern their own lives through them.

The main target of ironies, in the novel and in one of the proposed short stories, is evidently the optimism of Gottfried Wilhelm Leibniz who in *Essays on Theodicy* spoke *about the goodness of God, the freedom of man and the origin of evil*, so I bring to this work Essays *on Theodicy* that defines *'optimum'* to provide a greater understanding of what is narrated and the modification in the semantic load of *'Optimum'*, which is the same as 'Best Possible', very worked on in two of the proposed fictions.

Leibniz was chosen, obviously, for optimism and not for religion or metaphysics, which is proven by the author's expounded view of science and faith in Letters to M. Saint-Cyr of the French Academy, February 1743, where it is possible to read:

"I have the honor to send you the first pages of a second edition of Newton's Elements of Philosophy , **in which I present an outline of his metaphysics. You will see that Newton, of all philosophers, was the most convinced of the existence of God, and that I was right in saying that while a catechist proclaims God to children, a Newton demonstrates Him to the wise.**

I heard from you, as a great consolation, that I had dared to paint religion in its own colors in Henriade, and that I had even had the good fortune to express dogma **as correctly as I had been sensitive in the praise of virtue.**

I wrote against the fanaticism that spreads so much rancor in society (a kind of pharisaism among Catholic traitors) and that, in the political state, has stirred up so much disorder. But the more hostile I feel towards the **spirit of faction** (condemnatory religious partisan passion), of enthusiasm, of rebellion, the more I adore **a religion** (the real Catholicism is another and is followed by it) **whose morality makes the human race a family and whose practice is inspired by indulgence and good deeds.**

Stoicism has given us only one Epictetus, and Christian philosophy has formed thousands of Epictetes who are ignorant of being Epictetus, and whose virtue is even ignorant of virtue itself. It sustains us above all in misfortune.

I recognize that it was not genuine respect for the Christian religion that

> inspired me (Catholic) never to do any work against modesty; this must be attributed to the natural detachment that I have maintained since childhood in relation to such frivolities, these indecencies adorned with rhymes that, by the subject, please an unbridled youth (In reference to inconsequential youth.). At the age of nineteen, I wrote a tragedy inspired by Sophocles, in which there was not even love."

There is a clear tendency to criticize the excesses of traitors (fanatics) in relating God to faith and metaphysics to misrepresentations. Then his criticism of the young metaphysician will be intelligible, not because he dedicates himself to the study of metaphysics, but because he is wrong to define his own error as metaphysics, just as fanatics do with religion, which shows Voltaire's resentment clearly in an understandable way as a human characteristic.

I believe that the most striking characteristic of Voltaire is sincerity and I trust those who assume their own mistakes and Voltaire never hid his character.

With the proposed elements and others used as support, I think it is possible to draw a true, objective and focused study on the phenomenon that brings interpretative richness as a consequence of its evidence. The study of the rhetorical resource irony can be deepened when there is a greater understanding of the phenomena addressed when making use of it, so a complex study must always be accompanied by contextualized categorized referents and this work aims exactly to indicate the correct path for the development of an assertive interpretative reading, by bringing part of the description of the phenomena involved in the narratives necessary for understanding in a more complete way and by providing the Aristotelian scientific method, when necessary, to contribute to learning.

Doubts indicate the existence of caution and truth in speech in this reality, so I advance that I cannot conclude the truth about certain semantic modifications that require consultation with the author Voltaire, the only one who could give us certainty, which can demonstrate recognition of limitations and sincerity when speaking, however, obviously I will clarify the possibilities when necessary.

Studying the rhetorical resource, as this writing proposes, helps to read artistic texts and allows a higher level of understanding that allows one to appreciate the beauty of more complex literary texts such as the proposed corpus and, consequently, to read better. Certainly, this study is capable of instructing the reader by offering him the necessary ability to detect mistakes or rhetorical inadequacies more easily, which makes it necessary in the bases of literary studies.

In this, the universal analytical category is Socratic irony, the phenomenon of irony is limited to Socrates. The breadth of the definition that is divided into many ways such as romantic, tragic, comic, stable and unstable, of character, of plot, all based on verbal, situational amplitudes, in fact that such amplitude is observed as an accessory phenomenon, of excess of definition, nevertheless I bring the essence of semantic and dramatic inversions of support as well observed by Muecke, D. C., visionary of this essence.

This writing is structured by evidencing the phenomenon, by substantiating the definitions of the rhetorical resource, by bringing pertinent narrative aspects, by contextualizing the phenomenon in history so that there is an understanding and perception of its development present in the proposed works, by making critical analyses, by bringing conclusions that recognize the complexity of the phenomenon studied, by aiming to confirm the importance of the in-depth study defended and by discussing the evidence of the truth.

Jonas Batista dos Santos

II - ANCIENT GREEK IRONY

In Plato's *Euthydemus* (ΠΛΆΤΩΝ) text, Socrates (ΣΩΚΡΑΤΗΣ) reports to Crito (ΚΡΙΤΩΝ) who the brothers Euthydemus and Dionysodorus (ΔΙΟΝΥΣΌΔΩΡΟΣ) are and what they do. Socrates was aware that the brothers Euthydemus and Dionysodorus were experts in refuting, and possibly possessed the idea that refuting was one of the consequences of wisdom, and yet, forCrito, at first, Socrates recognizes that it is possible to speak falsely with truth, that refuting may not demonstrate true wisdom, and He refers to the brothers as wise and fighters in everything. Later, Socrates recognizes for Crito that the brothers' knowledge of the verbal dispute is recent, as it is possible to see:

> "It is that these two themselves, being, so to speak, old, have initiated themselves into this knowledge that I desire to eristic; last year, or the year before, were not yet wise." (Plato, *Euthydemus*, p. 35)

Understanding that they were new to developing their own eristics and wanting to know about their knowledge, Socrates is not afraid to enter into a dispute with his brothers.

What Socrates knows, at first, is that the brothers understood war, that is, they understood tactics, commandos and fights with weapons and that they also understood defense against injuries in the courts.

In his account to Crito, Socrates reported the past occupation of the brothers to Clenyas (ΚΛΕΙΝΊΑΣ) and was corrected by the brothers who, with contempt, said that they were at present engaged in teaching virtue, excellence.

Socrates also talks about Ctesippus (ΚΤΗΣΙΠΠΟΣ) who, enthusiastically, enters into a dispute with the brothers to quickly assimilate their technique and is very jokingly ridiculed.

Obviously, Plato works with the idea that the development of defense against supposed injuries made the brothers develop the art of refutation in some way.

The rapid transition from the occupation of the brothers to the teaching of virtue is remarkable, which shows that they had little time of experience and that, possibly, Socrates did not believe that the teaching of virtue, through the brothers, was something very certain, however, Socrates lets it be understood that he recognizes the importance of observing the beauty of phenomena as they are, that is, one does not ignore those who want to teach good things and I believe that this is one of the main messages of Plato in *Euthydemus* that converges in the posture of the wisest man in Greece who showed that he wanted to obtain a sample of the eristics of the brothers who considered themselves experts in everything.

Not ignoring serves to highlight the truth, that is, to expose the good or the bad in the occupations of others, it is not about giving importance to the good or bad occupations.

Refuting and teaching virtue are distinct phenomena and Socrates was only interested in the eristics of the brothers because he understood the teaching of virtue provided by them as false or ineffective. He asked for a demonstration of the new skill and gave signs that he doubted the effectiveness of the skill as can be seen:

The Rhetorical Resource Irony

> "Could it be that the one who is already convinced that it is necessary to learn from you, only those who are capable of making a man good, or also the one who is not yet convinced, because he does not entirely believe that this thing, virtue, is something that can be learned, or that you are masters of it?
> Say: is it the task of the same art to persuade such a man both that virtue is something that is taught, and that you are the ones from whom one would learn best?" (Plato, *Euthydemus*, p. 43)

In the questioning exposed, which has some offensive charge, there is the intention to induce, cause effect or change the thought of one interlocutor or more, which characterizes irony.

We have a sample of Socrates' masterful capacity for persuasion, who uses the rhetorical resource of irony to question the brothers by inducing them to answer whether only the convinced are capable of learning from them and allowing Socrates to question them freely.

In the ironic questioning exposed, sarcasm is present and has a certain and detectable place because it is supported by a system or *game* that runs outside of current and real life, as well as that which has the characteristic of being playful.

Through sarcastic irony, in the space that does not belong to ordinary and real life, offensive expressions are formed such as: "'Possibly you do not have the capacity disclosed.', 'I doubt your knowledge.', 'Those who learn from you are self-deceived.' 'I doubt whether you are the best to teach virtue.'"

Obviously, the unspoken possibilities raised are the fruits of sarcasm, which lead the interlocutor to solve them and assume the truth about himself.

Questioning whether only the convinced are capable of learning from them induces them to allow Socrates to question them freely.

Clearly Socrates was very interested in the eristics of the brothers and, in order to acquire a sample of it, he questioned to get them to give him a sample.

Dionisodorus confirms that it is, but it is not the confirmation that Socrates wants. Socrates wants the demonstration and, through questioning, led the brothers to demonstrate.

Because he was more experienced, Socrates continued to use the same rhetorical device to induce the brothers new to eristics to have some reaction, as it is possible to see with the following questioning when pretending to want an answer:

> "Are you, then, Dionysodorus," said I, "of the men of today, the ones who best exhort you to philosophy and the cultivation of virtue?" (Plato, *Euthydemus*, p. 43)

Socrates then told the brothers to leave the great demonstration for later, and that at that moment they should persuade the young Clinias that it was necessary to love wisdom.

Socrates gave a great responsibility to the brothers while pretending that it was a lesser responsibility by referring to the demonstration of what he did not yet know as greater, that is, while Socrates asked for a sample of persuasion, he persuaded or led the brothers to do what he wanted, to give him a sample of their eristics.

Socrates, without informing that what he wished to receive from the brothers was a sample of eristics, received a sample of eristics, which can be characterized as ironic.

It is important to note that, for there to be irony, there must be intention, because if there is not, the fact that someone provides an eristic sample without knowing that he provides it or provides something that he does not want to provide, as someone who has been deceived, does not characterize irony, which, in principle, limits the occurrence of what is ironic to the need for two or more consciousnesses to be one of them the inducting questioner and the other the one led to something, which shows that there is also the need to simulate in the sense of pretending and, because it is true that pretending is within a universe that may have a different time and place from current and real life, playfulness is always present in some way.

Obviously, in order to convince Clinias, Euthydemus and Dionysodorus had to offer him an eristic sample that would lead him to develop virtue and excellence, and Socrates, who was interested in the eristics of his brothers, led them to do what he wanted while worrying them by imposing on them the condition of being good at what they did in order to be able to teach.

In this type of questioning, where there is pretense when desiring an answer, there is the rhetorical resource, irony, in a useful use, and it served well, because, if Socrates had asked for a sample of eristics, he would possibly not have received it or possibly would have received it in an insincere way.

Much could be said about the nature and especially the honesty of Socrates in making use of many insults and hiding his real intentions, however, his actions are completely justifiable, as they are necessary reactions against the deception that insulted him. Rude language can be a weapon against the intention to deceive and we should not respect certain postures as well as we should also act with prudence in truth that irony can also serve to respond or return insults in a refined way.

Led to do what Socrates wanted, Euthydemus asks Clinias who, among men, are the ones who learn (*manthánontes*). Whether it was the "intelligent / those who know" (*sophoi*) or "the stupid / the ignorant" (*amatheîs*).

A lie is always supported by some truth, because there is no absolute lie. One lies to deceive and to deceive some level of truth is needed to support what is concealed.

If the lie had an observable body, it would have many zones of intersection between deception and truth in the most perceptible parts, and both answers proposed by Euthydemus touch the field of error and success, depending on the circumstances, because the spoken and written truth is dependent on the true circumstances.

This happens because spoken and written truth is an adjustment between linguistic signs, semantic loads, referents and circumstances, which leads cheaters, liars and those who wish to pretend to be right to invent inadequacies in one or more elements used to express the truth.

The main element modified by not conforming to the truth or lying is the circumstance, the condition

The Rhetorical Resource Irony

that involves time, place and manner, which allows us to say that something is true in some way, that is, under certain masses and forces involved that may or may not be real, concerning whether it is real or not determining whether it is true or not. Linguistically, the greatest occurrence of modification that allows not to conform to the truth or lie would be in the terms associated with verbs and in the semantic loads of nouns.

In the questioning to Clinias, the circumstances in which the intelligent and the stupid learn are not informed and it is true that the intelligent and the stupid can learn in many circumstances, because it is true that those who know can know more about what they already know or learn something new and the ignorant can learn something at some level.

Socrates, for example, when referring to his learning, never took wisdom itself as a complete phenomenon.

By referring in this way, if there is no observance of the inadequacy present in the circumstance of the questioning, it is possible to discourse within the deception of the questioning as if there were a search for the truth, when in fact it is a search for a non-existent reason in a null limitation stipulated by the questioning.

Dionysodorus, with a broad smile, spoke in Socrates' ears that any answer given by Clinias would be refuted.

With this, Socrates understood that the knowledge demonstrated was an initiation game or was a technique of the brothers given to refutation used to deceive.

With the clear intention of ridiculing the brothers, if their eristics were supported by rhetorical inadequacies, to encourage Clinias to continue, Socrates persuaded him by telling him, possibly without being sure, that these games were a kind of initiation, which may show that Socrates cared little about Clinias' emotional state in the face of his learning and his observation of eristics. a posture of someone who understands that a man must go through afflictions to grow.

Socrates demonstrates that he understands and respects the conditions of interpretive happiness in the real world to the point of being completely detailed in considering all the possibilities to undo the confusion that served, at first, as a technique to deceive or to bring confusion.

Obviously, if a question does not have a context in which it can generate a completely assertive answer, when questioned, in self-defense, the brothers should defend the idea that no one makes mistakes in any way and fall into contradiction for that.

In response to the brothers' assertion that no one makes mistakes in thinking, speaking, and acting, Socrates, who detected the contradiction that served for self-defense, raised the following ironic question against the brothers:

> "If we do not err, neither in acting, nor speaking, nor in thinking, you, through Zeus!, if so, have you come here as teachers of what? Or is it not true that you have just said that virtue, better than any other of men, you can transmit to those who are willing to learn?" (Plato, *Euthydemus*, p. 77)

Often, dealing with a matter with truth is not of considerable effectiveness when one is being deceived, when the interlocutor is willing to lie in order to advocate for one's own cause. In this case, irony has much more persuasive power to bring the truth about some phenomenon than exposing the truth directly.

Faced with a clearly unreasonable insult, it is better to question why it exists for those who uttered it than to return a worse one to those who are not willing to speak truthfully. Let's look at the following example that shows irony converging in questions that pretend to consider the answers:

> "- This man is a dog!
> - Before I tell you anything, I ask you: Why do you consider him a dog?
> - Dogs are incapable of refuting me.
> - I think that your thought is disassociated from reality. Can you cite an example? Can you better describe the deserving of the offended man?
> - Dogs bark unnecessarily and so are all those who study in this Lyceum.
> "Do you have nothing against the offended man, and divide those who study at the Lyceum into a kennel for no reason at all?"
> - You're a dog too. The kennel is the Lyceum. Shut up you too, dog.
> - All of us who question stupidity are dogs and you, who were offended for being stupid, would be what? I ask to understand the scope of your madness."

It is true that the one who affirms does not want to be contradicted and we note that the mind of the interlocutor works as if he disregarded himself as the target of the offenses, because he does not perceive the truth in what is said against him, so he developed the ironic question "- Do you have nothing against the offended man and divide those who study at the Lyceum inside a kennel for no reason?" uttered by an interlocutor who pretends to want an answer and who, by acting with dissimulation, possibly seeks the reflection and wisdom of the other so that he sees the truth and even his own moral misery without offending him, at first.

This is the question of a man who is above everything, who may not even feel enough benevolence to give an offensive answer to educate and who only observes an animal in its state of emotional uncontrol.

When the interlocutor asks if the other could better describe the deserving of the offended man, namely that the offended person did not deserve the offense, he is forcing the offender to recognize his own stupidity even if only internally.

Let us note that the questioner knew that the offense had no reason to exist, so by pretending to want answers, he led the offender to reflect on his speech, to understand that he offended without reason.

Another way to use the rhetorical resource irony would be, in this case, to give the theoretical basis for what it is to be a dog while creating disassociation between the primary offended party and the definition and directly offending the sender with the definition. To do so, after giving the definition, it was enough to question who would be more appropriate to the definition, who would be offending in a similar way to barking for no reason.

Regardless of the effect, irony exists in questioning, to serve as a rhetorical resource in the face of an unreasonable offense whose origin may be indefinable.

Irony, when used for the sake of truth, has the characteristic of exposing the deception of the deceiver's interior, who starts to feel indignant and ashamed by the exposure of himself, so in the proposed corpus, when caught in contradiction, the brothers do not answer Socrates' ironic question and Dionysodorus offends him by insulting him as an old gaga.

This is the technique that all the bad guys use for self-defense. They pretend to be victims to justify their own evil.

I say that the brothers' technique consisted of observing opposition to what is said in conditions that favor contradiction, that is, they created rhetorical inadequacies to be right and Socrates understood this when he said:

> "Men like you, while others are so ignorant of them, that I know they would be ashamed to be ashamed of themselves before refuting others with such arguments, than to be refuted themselves.
> In your arguments: when you say that no thing is beautiful, or good, or white, or anything of that kind, and that it is not at all other than others, you simply, in reality, sew up the mouths of men, as you also say; but, because not only those of others, but it seemed that your own as well, this is very gentle, and takes away what is unbearable in the arguments.
> But what is more important is that these things which you do are such and how these arts are invented by you, that in a

> very short time any man could learn them. I realized this myself, paying attention to Ctesippus, how quickly he was able to imitate you immediately." (Plato, *Euthydemus*, p. 127)

This reveals that the initial impression about the time of experience of the brothers evidenced by Socrates was correct.

They had quickly developed a technique of deception that consisted of creating inadequacy in one or more of the elements of happiness necessary for the truth to be evidenced, that is, the inadequacy would be in the linguistic signs, in the semantic loads, in the referents or in the circumstances.

In fact, what is beautiful requires time and care, and probably Socrates understood this. The eristic technique of Euthydemus and Dionysodorus was new and ignored the complexity of beautiful phenomena, a technique that could be acquired so quickly that Socrates, ironically, recommended that they restrict the use to only among themselves so that others would not learn quickly and refute them.

To do so, Socrates used the example of Ctesippus who quickly learned the technique from his brothers. Let's look at the moment when Ctesippus realizes the inadequacy:

> "It's just that he says things that are a certain way, but not as they are." (Plato, *Euthydemus*, p. 69)

They are in a certain way, that is, in a certain circumstance. Ctesipo realized that the inadequacy, in this case, was in the circumstance, that is, linguistically, they did not use terms associated with verbs to hide the context and use the inappropriate context to the answer of an interlocutor to always be right and win any dispute.

Let us observe that, in a rhetorically appropriate way, when we use an adverb that opens up appropriate possibilities, we can reference referents whose complexity of beginning and end are indefinable by words, but such adequacies are not desired by those who wish to cheat.

Article	Noun	Verb	Adjective	Adverb
O	**Sky**	**is/becomes**	**blue**	**periodically.**
Noun Adjunct	Subject	Linking Verb	Subject Predicate	Adverbial adjunct

To be more specific, in fact that the sky does not only have the color blue, we would add the adverbs **'predominantly'** and **'periodically'** after the predicate of the subject **'blue'** and, finally, to limit it even more, the adverbial phrase **'in our perspective'**; **'The sky is predominantly blue, periodically, from our perspective.'**.

In this case, the adverb **'predominantly'** takes into account other colors in less predominance, **'periodically'** indicates that the color may not be predominant or almost non-existent depending on the context and **'in our perspective'** indicates that the perspective is human and not from another verifiable reality.

When affirming and questioning, we must take into account these descriptive happiness rules, so that we avoid falling victim to arumentative cheating.

Let us note the many interpretations when we do not restrict the senses to real possibilities:

"- Is the sky blue?

> \- Yes.
> a) - No. You see blue, but it is colorless.
> b) - No. It is blue at night and at the end of the day it does not have a predominantly blue color.
> c) Yes and no."

I believe that 'yes and no' are extreme answers with a certain level of contempt for the subject.

When we are asked by a cheater about any topic, we give the context along with the answer:

> "- Is the sky blue?
> - During the day, its color can vary in predominance of blue, gray and white."

It was used to open possibilities ('during' 'could be'), a list of restrictive terms to them ('blue, gray or white.') and never would disputants like Euthydemus and Dionysodorus have such honesty in speech when disputing to obtain victory.

For Crito, concerned with the education of children and with those who make mediocrity an occupation, Socrates says that, even if there are many mediocre people occupying positions that they should not, it is not fair to ignore those who teach well, that is, it is not because there is a bad example that we should judge all others by it. In fact, there is, in many others, the notorious value of aversion to liars and deceivers, those who abhor rhetorical inadequacies such as those exposed by Euthydemus and Dionysodorus, which are the property of part of those who make use of techniques to deceive.

It is evident that the Greek text can have an effect on its readers by disseminating values through severe criticism of deception, and its intention to educate by fostering the search for virtue is clear.

The Greek word 'ειρωνεία', which originated the word 'eironeia' in Latin and which originated the word irony in our language, refers to dissimulating when questioning and Socrates provides us with a sample of irony that serves to highlight the truth and find virtue, excellence.

Irony can manifest itself in Greek disputes, when raising questions that can lead an interlocutor or opponent to take some action, to be contradicted by their own intellect and to reflect on their own ideas that may change, that is, the questions can lead an interlocutor to contradict himself with his own argument, whether his answer is verbalized or only mentally constructed when completed, even without verbalizing, that the phenomena are not as previously concluded.

Obviously, for there to be dissimulation, it is necessary for two or more consciousnesses to interact, which greatly restricts the phenomenon of irony as a rhetorical resource. The same may not happen with the text that proposes to be ironic without direct interaction between two or more characters.

As for success, it is not necessary to convince for the phenomenon to materialize. Success lies in not being deceived.

For irony to exist as a resource, it is necessary to have a situation that allows the use, that is, it is necessary that two or more interlocutors who do not agree with an idea or are unaware of it, that is, there must be speech or expression, response, disagreement of idea and search for some conclusion, which does not prevent two interlocutors from referring to the ideas of a third party who is not present, because the result of irony does not contribute to its existence as a primary cause, because there is no true irony for the sake of irony, there is a need for the resource to exist for the sake of another phenomenon and, once the phenomenon is developed as a resource, its condition of happiness in effectiveness is not guaranteed when trying to convince.

What contributes to the emergence of irony is what motivates to cause the effects generated by the phenomenon, to lead to promote actions, which includes changing what the other thinks, which can affect even those who witness without participating.

Obviously, this property of having an effect on those who do not participate allows the creation of ironic texts that they believe will have an effect on readers and, possibly, it was what allowed the broadening of the definition of 'irony', since the revealing power of irony can be present in many artistic texts in many different ways.

It is important to understand that not every artifice that can generate some effect on interlocutors that leads to a change of opinion due to contradiction is irony.

In addition to the situation that allows the phenomenon to exist, there must be conduct that has the motive of leading an interlocutor or more than one, through questioning, to do something or to have an idea different from the one he has that may be contrary, because it is in dispute, that is, the interlocutor is induced to do or conclude something that contradicts him in a conscious state or not. The interlocutor may even contradict himself as an effect after being the victim of ironic questioning, as happened with Euthydemus and Dionysodorus.

The Rhetorical Resource Irony

Irony can transcend defined limits, but all its characteristics must be maintained in some way. The questioning must exist in some way to be ironic, even if in a presupposed[1] or provoked way[2].

A reader may suffer the effects of irony when reading '*Maybe I don't know how to answer why I love you so much.*' within a context by creating many questions through the answer, in observance that answering mentally indicates that the question '*Why does he love you so much?*' exists somehow.

In this case, without the context, it is impossible to determine whether the question that does not exist graphically in the text is ironic or not. If the answer is '*He loves you because he considers it the right thing to do.*', for example, it may or may not be ironic depending on the context.

In a context where the gentleman is forced to marry because of his high values that oblige him to love his lady without actually recognizing attractive qualities in her, and conclusions are produced that indicate that he was forced to marry and love his lady even after she became horribly ugly[3], so that it changes the reader's thinking that he considered that he loved her very much for her qualities or believed In the lady's aptitudes of physical and spiritual origins, there is irony in some way.

This happens because, when defining the context, the contradiction between the speech that was taken as a result of satisfaction and the situation exposed

---

[1] To exist in a presupposed way is equivalent to being essential for the existence of something.

[2] To exist in an entailed way is equivalent to being part of a logical sequence as a possible consequence.

[3] The example of Candide's Kunigunda was used, in part, in *Candide or* Voltaire's Optimism because she had become ugly and Candide had married unwillingly out of pity.

later, which does not allow one to take the attitude of marrying for a pleasure but for an obligation, becomes evident.

Obviously, what is offensive works for the playful, which can provoke the anger of many. Voltaire, for example, was arrested more than once for his literary productions and ironic statements that should not be taken as offenses but as literary art or genius.

Satirizing may be a form of homage, but many, without having developed a high understanding, react against satire as if the elements of it were real-life. This is what happened with Philip d'Orleans, who provided imprisonment for Voltaire, who was accused of satirizing his intimate life, the intimate life of the spoiled son of the King of France.

Ironic questioning, which works for the playful only, is supported by the offense that may be real. In most playful occurrences, it exists in a presupposed or entailed way and as an essential resource.

Questioning by pretending not to know the answer can manifest itself in many ways, but it has only two moral paths that are distinct.

There is irony when one of the interlocutors, with dissimulation, raises a question that forces the other to contradict himself or to have an opinion different from the one he had, in case two affirm differently without reaching a consensus.

When this lead is to a truth, there is usually a questioning that respects the correct use of linguistic signs, semantic loads, and referents in a given situation in the world, which makes it possible to reflect on the truth if all possibilities are taken into account so that the truth is refined.

A question does not have, in itself, the characteristic of being adequate, because being true or being inadequate are particularities of individuals who flow through their speeches, so there are questions that lead to the truth and those that lead to deception, however, it is not possible to record them to be identified completely, because they are dependent on circumstances and the same linguistic signs can have different semantic loads depending on the circumstances.

As seen in the example of Socrates, sarcasm can also be used in leading to the truth and is an element present in many ironies. From the Greek 'σαρκασμὸ' (sarcasm) which means to mock, sarcasm is what carries the necessary offenses for sarcastic irony to exist.

When questioning or answering with respect is condoning deception, sarcasm can infuse its victims with part of its deception to contribute to the desired effect, discovering that the source of the error is the sender himself, if it is used appropriately to the truth.

Sarcasm is very useful for the evidence of the truth when used in argumentative disputes in an appropriate way, as it has the function of giving reason, in part, by mocking the dissertation or the knowledge of others indirectly, that is, it brings a previous conclusion of defeat for the onslaught of a deceiving opponent.

Often cheating can be built in two moments, so cheating people usually criticize the criticism of the use of theoretical bases or elements prior to the current speech:

"- You could lend me a certain item.

- Yes. You can use it to go to meet you, for example, and then return it to my garage.
Thank you. When I need it, I'll let you know.

...

- Friend, I'll need it today.
- You can take it then.
- When I organize myself on certain issues, I'll give it back to you. I will take good care of what is yours.
- How can you be so honest with me? On second thought, I decided not to lend you any more.
- I don't understand.
- Yes, you understood and planned very well.
In what we agreed it was implied that you would bring it on the same day, but you changed the agreement, you did not define the return. Why did you break our agreement? Are you immoral? Unfaithful? Cheater? Liar?
- You have no word and you are not my friend.
- Of course I'm your friend. I'm helping you with what you want. Was it not you who broke our agreement?
When we woke up, did you already plan to cheat or did you decide to cheat after the deal?
- Make good use of it. God bless you.
"I will." And this inner person of yours that interests me, I am interested in knowing if you planned the cheating before or after and all that this question entails. I already know your considerations and they are very

> common. I want to understand to learn more." (Part of actual transcribed dialogue of straightforward sarcastic irony.)

With the use of sarcastic irony, all deception is redirected to its source, the deceiving interlocutor who mentally tastes his own vomit.

There are perverse argumentative manifestations in the sense of being adverse to the truth, so perverse that only mockery can promote an adequate response and it is for them that sarcasm should be used, however, obviously those who love deception would have to make use of sarcasm to pervert by making inappropriate use to promote deception by mocking with false representation of the truth, by creating a false image and liable to defamation of their opponents, of some value or some given definition.

Sarcasm is also used assertively when there is no certainty that one is being the victim of bad arguments because it allows questioning, however it is positioned very close to directly offending that usually ends in wasting time with silly speeches so that time can be better spent on important issues.

Notwithstanding what I say about the use of sarcasm, silence when being the victim of insults is the best way and initial resource. When being accused, if we are forced to answer, we can just say "You say so." and let the accuser prove it. Then, by receiving all the false accusations about you, we can ironically question your origins without the foolish pretense of getting recognition about being right.

Socrates dared to say that the art of the brothers consisted of a technique of deception that served only to sweep opponents away, however, he wanted to have a sample of everything they could offer.

Ancient Greek irony can lead the interlocutor's reasoning to be inverted or mocked and, in the case exposed, the ironic questioning served, later and in part, to discredit him, which allows us to say that a sample of very complex sarcastic irony was also obtained, which is difficult to detect.

Socrates' attitude should never be confused with the mockery against the truth that is very popular among those who do not know what they are saying, among those who speak with inadequacies to be right.

In the fragment exposed, Socrates dissimulates as in the Greek definition evidenced and raises questions, but in fact, he does not want to obtain the answers to the questions. He is doubting, in secret (dissimulating), with *sarkasmos,* what the sophistic brothers who arrived in Athens with the reputation of illustrious wise disputants capable of making virtuous people through persuasion or teaching say.

In the corpus studied, Socrates does not believe that the sophists Euthydemus and Dionysodorus can teach virtue, because the acquisition of it is not certain when acquiring knowledge and he does not believe that they can have the qualities of masters that they disseminated in Athens, so he developed questions to make use of the rhetorical resource irony, but there was a little good expectation in Socrates. He was curious and wanted to have a sample of knowledge, he obtained and concluded, at first, that that eristic consisted of a technique of deception that could be learned quickly.

In the text, the foreigners Euthydemus and Dionysodorus say they have all the science from the cradle, a knowledge capable of knowing everything in the world, including the number of teeth of their opponents present in the dispute, which are, at first, Socrates, the young Clinias and Ctesippus.

Obviously this is a reference to the techniques of detecting expressions and prejudgments by appearance and obviously knowledge is not ready in any mind as it is understood when it is understood when it is said that one has all knowledge, because it must be processed and Socrates, obviously, doubted the qualities exposed, but believed in the beauty of the dispute, even when being the victim of arguments that he believed to be part of a technique that only serves to deceive and that does not it represents true knowledge.

In the inability to show true knowledge, the brothers offended Socrates and his followers without having reason to do so, which does not work against the existence of humor in favor of the evidence of the truth, because, as has been said, by the nature of irony, whenever the phenomenon develops, there is the ludic in its own time and place that is not current or real life.

Humor is always offensive, humor always ridicules idiocies, imbecilities and, at worst, the decadence of others.

For example, those who laugh at situations of nudity and bestiality are, in fact, laughing at the people who are being exposed to the aforementioned elements, that is, they are laughing at the offended and not at the elements of nudity and bestiality, which indicates that there is pleasure in individuals, possibly degenerate, who develop this supposed humor when they perceive the decadence of others.

In a humorous presentation without words, it is possible to laugh when the presenter pretends to be deceived, for example. Many people like the feeling of superiority in relation to other people and this type of humor provides this.

Laughing at idiocies and imbecilities is useful in educating, which is very different from laughing at the decadence of others. It is possible that there is disassociation between individuals and their acts when highlighting customs in playful theatrical performances, for example. Offending people playfully by referencing them may often not be a real offense. For example, by repeating a derogatory lie about an individual or place whose verification that what is said is false and easy to acquire, there may be a clear intention to provoke laughter or re-attest to the values contrary to the repeated falsehoods. In other words, ironically offending may have the objective of paying homage from within a playful situation to reality. Voltaire used this a lot when attributing behavioral patterns to individuals who did not have them or hid them.

To say, *"Their beaches are horrible, and they don't disdain ours for that reason."* When it is true, verifiable that the beaches are beautiful, it is a tribute to the beauty of the beaches and to the common character of the citizens of somewhere.

If their beaches are beautiful, the reason not to disdain them would be humility. That is why there is a playful depreciation of the character of the common citizen of some place that generates homage.

Let us observe these two statements. In one of them the supposed power exists in some form and in the other it is only cited to mock for sarcastic irony:

> The supposed power manifested itself **as** some disorderly movements.

> The supposed power was manifested **by** some disorderly movements.

In the second statement, the supposed power can exist in some way through some disordered movements, that is, they are distinct phenomena and dependent power is acquired through disordered movements. It turns out that, if this is true, there is no irony in stating such a thing.

In the first statement, disordered movements form the manifestation of power and, while it is true that, by following the Aristotelian rule on substances that says that what flows from substances is what can predicate it, since 'power' is an inadequate predicate to define disordered movements, the sarcastic irony, the mockery and derision against the phenomenon *of disordered movements is evident.*

A person who is inattentive when looking for logic about substances and their possible predicatives may not perceive the rhetorical resource of irony in this case.

Just as in a conversation between two interlocutors where one questions by pretending not to know the answer, the one who preaches with sarcasm when seeking to evidence the truth can lead the interlocutor or the reader to question their own convictions, in this specific case, if the disordered movement forms the completeness of a supposed power, it leads to the self-questioning of the existence of the supposed power manifested by the acts themselves, which shows the presence of sarcasm.

Faced with the offense of the brothers Euthydemus and Dionysodorus, who considered that their interlocutors were incapable of detecting their fallacies, Socrates addressed them as *sarkasmós*. Even

Socrates inducing foreigners to fall into contradiction, so that he caused those present to almost die of laughter, questioned why they laughed at something so beautiful.

Socrates did not perceive beauty in deception. He perceived beauty in the questioning of deception that could allow him to add to his own intellect samples of how not to be. The moment itself was beautiful.

Socrates perceived the beauty of the argumentative dispute, in the eristic, even if deceptive for its usefulness as a sample, and those present, Clinias and Ctesippus, should feel great emotion before the beauty and power of the truth brought through the questioning of the wisest man in Greece, but, by the nature of the phenomenon and by the human nature that leads to laugh with satisfaction at a certain level of human decadence, according to Socrates, they almost died of laughter after he exposed a contradiction in the brothers' speech.

Sharing the same perspective as my master Socrates, I made use of the resource in a notorious episode after receiving an invitation to participate in an oratory event.

At the event I heard about how to speak well in public and, dissatisfied with having been exposed to a routine of adulation and cheating, when they started to ask something to the speaker and his associates, I raised my hand feeling insulted and decided not to enroll in the course offered by the speaker.

> "- Introduce yourself. Say your name and what you do.
> - My name is Jonas and I live in the countryside, in Boqueirão dos Coxos and I work in the fields.
> - Right. And what is your question?

The Rhetorical Resource Irony

- In order for me to speak well and truthfully about a specific subject, shouldn't I study it deeply, in fact that the proposal is to speak well?
- Yes. I know where you want to go. Is Jonas your name? I didn't hear well.
- That's right. Jonas from the countryside. The doctor speaks well about treatments and uses of medications.
- Not necessarily. A person can study and speak well too.

(At this point, some in the audience applauded the speaker.)

- So you have admitted that specific study is necessary?
- Not necessarily.

"So you offer the ability to talk about everything?" What does public speaking consist of?
- Not over everything. To speak well is to express yourself fluidly.
- What are you a teacher of what specifically?
- Rhetoric, personal marketing...
- To sell one's own image, whatever it may be?
- No. The person has to have something to offer. You, for example, work with what?
- In the countryside.
- You may be in the wrong place.
- But if I want to speak well in public. How do I do it?
- Talk about what if you don't have content?
- Exactly! From the horse's mouth! You assumed that in order for me to speak well about something, I need to know

and know in depth, because we all know little."

At that moment I was avoided and felt the bitter regret of the rejection of some, which is an illusion and also, in a high state of consciousness, I felt the inner pleasure of being unbearable in the right way with my consciousness aligned with the eternal, the truth.

The Playful and the Misleading

We cannot know about all phenomena at the same time in a complete way and Socrates, the wisest man in all of Greece, recognized his own limitations.

We all create clusters of information and concepts linked to linguistic signs, which can make us simplify and offer examples or linguistic signs linked to phenomena, instead of definitions, in many of the times when we offer quick answers.

When dealing with the big questions, using an example as a definition can be misleading, as is perceived in Plato's dialogue *Euthyphro,* where Socrates highlights the importance of the universal definition while Euthyphro uses himself as an example, the highest personification of justice.

We simplify to process less, that is, to have relief and the ludic serves as a superior relief of tensions according to Johannes Huizinga, author of *Homo Ludens*, and the ludic tends to omit and modify the senses.

At some point, we have to make use of some escape from reality so that we can have some mental relief that does not occur with the inactivity of the mind, it occurs in the practice of something different and working the opposite, the counteract, can promote playfulness and relief.

Since the description and discourse in the ludic is not about current and real life because it is different from the real, by the natural search for relief, I could comfortably define it as deception, but this would be a mistake, since simulating and exemplifying are necessary resources to offer more complete possible or real samples that cannot be obtained at the moment.

Only moral perversion can promote the motive to generate deception when lecturing and narrating.

Just as there is no phenomenon in the linguistic referents that form the linguistic signs referring to them, the perversion is in the motive and not in the written form, so it is necessary to know the intention to distinguish the playful and the deceptive.

Note that playful statements are more likely to be defined incorrectly than misleading statements. Thus, when Voltaire, in literature, jokingly addresses Gottfried Wilhelm Leibniz's theory about the world being the best possible, he does not define it correctly, he modifies its meaning, which is replaced by another that allows playfulness.

It is notable that the inadequacy may not be in the text, even if it is inadequate.

This type of playfulness only exists in literary texts that have much more than informing them by reason of existing.

Thus, it is not possible to consider a derogatory and isolated statement in a literary text, with no room for other interpretations, as deception or truth when the author's text is not clearly intended to depreciate, but the author whose intention is not completely exposed in the statement.

Certainly the rhetorical resource irony serves the jocularity, however the inverting of situations when the resource is used in its non-useful form for the evidence of the truth, that is, in its playful form, can promote deception in the most uneducated media, which is probable because there is the phenomenon of foolishness among the unwary who use unrealistic fictional elements as a basis to corroborate with some erroneous idea previously bought as truth.

An ignorant individual who has never read *Theodice's Essays* may conclude, after reading *Candide or Optimism*, that Leibniz is wrong or that Voltaire refuted and destroyed Leibniz's image, because he has assimilated the semantic charges of fiction that are inadequate in reality and because he does not have the proper capacity to interpret literary texts.

I understand as readers the people who have achieved analysis at the level of the organization of the phrases, but I understand that only benevolence in the heart leads to the truth, so it is not much to study what leads to having a good character and, consequently, good rhetoric. Good readers know little, define little and avoid presumption.

When ironic conduct serves the phenomena other than deception and playfulness, there is no respect for the adequacy of the elements necessary for there to be truth. Authors use the same source when modifying the semantic loads, however, they serve different phenomena and what is the result of playfulness serves the truth, because, if it were different from that, it would be perversion.

Certainly, in some cases, it can be very playful or facetious to modify the semantic loads of some words or contradict narrated situations.

The change in the meaning of words and the contradiction of a situation narrated in literary texts promoted by rhetorical inadequacies are phenomena evidenced by theorists of the irony phenomenon as literary ironies. Just as D. C. Muecke in *Irony and the Ironic* names, in literature, inadequacies or semantic modifications of verbal **or** instrumental **irony** and speeches of characters that contradict narrated situations of **situational irony**.

What theorists call **dramatic irony** occurs when the reader, in the case of literary works, knows more than one or more characters in a narrated situation.

In critical analyses it is noticeable that literary ironies have a common root, which will be proven later.

Irony is a rhetorical resource under the conditions exposed, however I have cited some types of irony described by modern theorists that seem to be distinct manifestations of a real case where an interlocutor makes use of the resource in a dispute.

The exposed resource, in addition to providing playfulness, makes it possible to provide deception, because the understanding of what is playful or true, where playfulness begins and ends, is not everyone's property, so I bring two more samples that exemplify the use of questioning to lead an interlocutor, in an argumentative dispute, to draw playful or misleading conclusions from their own perception through rhetorical inadequacy.

In these examples, Dionysodorus makes Ctesippus, one of the disputants who affronted the foreigners, conclude that he has a dog for a father and Dionysodorus' motive is to laugh at Ctesippus in order to win the dispute because he is right in some way and it is not noticeable the promotion of deception in these words but of playfulness, Although the reason shows that the deception exists in some form:

> "Tell me then, 'Do you have a dog?' "It's a bad thing," Ctesipo said. "Anyway, does he have puppies?" "Yes," he said, others like him, and a lot. "So their father is this dog, isn't he?" "Truly, I personally saw him covering the dog. "And then?" Is the dog yours? "Perfectly," he said. Isn't it,

> being a father, it is yours, so that it becomes your father the dog, and you the brother of the dogs." (Plato, *Euthydemus*, p. 113)

This is an example of using an inadequacy that allows you to speak falsely with truth. If the dog, which is father, belongs to Ctesippus, then the dog, which is father, is father and is his.

Certainly, there is a lack of semantic complementation and referent error to be able to speak falsely with truth. In the example, the ignoring of certain conditions of happiness in writing and interpretation creates a space for irony, playfulness and playfulness.

Despite the conversion from synthetic to analytical language, when Dionysodorus says 'being a father, it is yours', it is possible to perceive the rhetorical inadequacy that ignores the present logical condition that defines that if the dog were not a father, he would not be the father of Ctesippus. This inadequacy is manifested in the construction of the expressions.

In this other example, Dionysodorus makes Ctesippus, his opponent, admit that it is good to have gold in the belly, gold in the skull and gold in place of the eyes because he has agreed that it is good to have gold always everywhere:

> "And Euthydemus held his peace. But Dionysodorus, alluding to the answers given by Ctesippus earlier, asked, "Is it true, then, that gold also," he said, "seems good to you to possess?" "Of course," Ctesipo said, "it's a lot of him." "And then?" Do you not think that good things must always and everywhere be had? "Decidedly," he said. "Do you agree that

> gold is also a good thing?" "Yes, of course. I agreed, he said. "So, you have to have it always and everywhere and as much as possible about yourself?" And would it be very happy if he had three talents of gold in his belly, a talent in his skull, and a stater of gold in each eye? "Truly, Euthydemus," said Ctesippus, "it is said that among the Scythians the happiest and best men are those who have a lot of gold in their skulls, in their own, just as you have just said that the dog is my father; and, what is even more admirable, that they also drink from their own skulls adorned with gold, and contemplate them from within, holding in their hands the crown of their own head." (Plato, *Euthydemus*, p. 115)

Ctesippus, realizing the inadequacies that only served the jocularity and the loss of his reason, still tried to be right by falsely agreeing with Dionysodorus and by expounding on the use of gold within the situation exposed by him, which is an error that demonstrates that he respected what the brothers said when trying to imitate them and complement their reasoning.

If he had more experience like Socrates, he could have used sarcasm and shown that they were talking about different phenomena and that there were many inadequacies such as referent errors in speaking, but Ctesippus wanted to be right against a reason that only existed and worked in the ludic and that, in reality, was extremely inadequate.

This shows that any discourse that is actually contrary to a ludic phenomenon is null, because it is refuting something that does not have the objective of being true, something that is not even verifiable in current and real life, and the pact of verisimilitude requires that we only believe in ludic elements within space and time that do not belong to current and real life.

Possibly, Plato, the author of Euthydemus, worked on the caution of the characters regarding the offense against the brothers, because the direct offense in a dispute indicates loss against something that is not possible to achieve, which is perceived on the side of the brothers when they curse Socrates as an old gaga.

The intention of Dionysodorus' ironic questioning is to ridicule and, consequently, provoke laughter, which can lead to a series of favorable conditions for him, because an opponent demoralized in a single idea can generate distrust in other arguments of his and this is the intention of the mockery of the characters Euthydemus and Dionysodorus who won the dispute, according to the arbitrariness of fools, against Socrates and his followers without offering a sample of true knowledge.

Ctesippus referred to having a supply of gold in specific places as anyone with common sense could understand, but he acted wrong by respecting Dionysodorus who wanted to ridicule him, because his intention was not to be right but to win the dispute, so he modified the conditions of the referents that were mistakenly accepted by Ctesippus who started to argue within the proposal exposed to him, where the argumentative trap was clearly led due to the complexity of the phenomena involved.

It is difficult in a dialogue to shield the discourse against misinterpretations, because it is not possible to be fully aware of the phenomena involved at all times and in the perspectives of those who listen or read. From texts and speeches that express part of reality, when there is not the necessary knowledge to develop understanding, one can make mistakes.

Argumentative traps do not demonstrate knowledge and are, as Socrates said or as Plato wanted us to perceive, a technique for deception that overrides true knowledge.

In the work *Euthydemus*, we perceive two distinct sides, the side of Socrates that leads to the truth, that evidences the deceptions and that reaps the good even from the deceptions to find the truth, with the use of the rhetorical resource irony, and the side of Euthydemus and Dionysodorus who did not have the nobility of the search for the truth, therefore, they sought to win argumentative disputes by developing ironies through rhetorical inadequacies in favor of the playful or the joking, by attacking their opponents in order to win.

Unlike Dionysodorus, Euthydemus clearly had a different search and his speech contributed to the irony phenomenon having completeness, which leads to many other phenomena and conclusions about him, although they acted in collusion in some way. Euthydemus is the one who knows how to act correctly, but does not do it. At the right moment, Euthydemus assumes that he has fallen into contradiction, but resumes his routine soon after.

It is remarkable that Ctesippus sought to answer the brothers with his remarkable load of knowledge, but he lacked the use of any resource that would destroy their arguments, who, realizing that Ctesippus had great knowledge, tried to ridicule him. And Ctesippus could have used the rhetorical resource irony, which includes questions about linguistic logic about questions about himself.

The questions and arguments of the ancient Greek sophists that led to wrong or jocular conclusions taken as true are similar to those that Voltaire used in his great novel *Candide or Optimism* and in the proposed tales. This was because the Catholic Church had rescued Greek education and Voltaire had reaped its educational fruits.

Voltaire knew how to speak falsely with truth, knew how to lead the reader into argumentative traps, knew how to laugh and make people laugh in such a refined way that many, without proper education, considered many elements of his fictions as revelations of the truth.

Let's look at an example of a typically Socratic or ancient Greek ironic questioning of yours:

> "While you were hanged, dissected, beaten and rowed in the galleys, did you always think that everything was going as well as possible?" (*Candide or Optimism*, p. 128).

In the reality of the narrative, everything is as it is narrated in the exposed fragment, however, the semantic load of 'the best possible' is changed, that is, there is an inadequacy between linguistic signs, semantic load and, consequently, of referent and all of us, if we want to understand, must take the correct path to achieve understanding.

Because there is a semantic modification, that is, a modification of meaning, there is also a referent error, because the definition seen is not the same as that expressed by Leibniz in *Ensaios de Theodicea,* but the definition formed between the interlocutors and among many theorists of the phenomenon studied.

Semantic modifications such as those exposed contribute to produce playfulness (*game*) and are considered **verbal ironies**.

For Gottfried Wilhelm Leibniz, the best possible world holds afflictions and, for the character Candide, who learned the metaphysician-theologian-cosmo-lonigology from the preceptor Pangloss, at first, 'the best possible' does not hold afflictions, in a state of ignorance, until he witnesses them and concludes that lesser evils generate superior goods.

The exposed fragment of *Candide or Optimism* is from a part near the end of the great novel and shows a mature Candide who pretends to want to know an answer from his master, while leading him to reflect on his convictions and to draw conclusions contrary to them.

It is true that Candide intended to make his master Pangloss recognize that, in their practical life, phenomena are not as Leibniz says they are, and he questioned his master by pretending to want to know an answer to lead him to reflect on his own convictions, which characterizes his questioning as a complete manifestation of Socratic Greek irony.

If we take into account that there is the conscience of the author and the reader, we will know that the irony exposed affects the reader because it occurs between the author and the reader, between elements of the pact of verisimilitude of fiction and what the reader knows about a certain subject, making it possible to cause a change in the way the reader conceives the phenomena discussed, just as it was said that the phenomenon could influence non-participating listeners. In this case, there may be a transcendence of the playful to the real.

When Dionysodorus smiled broadly and spoke in Socrates' ear, he informed us what Clinyas did not know, which characterizes **dramatic irony** in some way, because we know more than a narrated interlocutor. What is exposed also puts us close to Socrates in terms of perspective. Here is a narrative technique to develop empathy in us readers for a certain element. That is why even fools praise Socrates, create scarecrows of his own so that Socrates becomes close to them.

Obviously, the phenomenon **of dramatic irony** exists in some form when the characters in the novel *Candide or Optimism* talk inappropriately about some phenomena. Let's see how this presents itself:

> "It is demonstrated," he said, "that things cannot be otherwise: for everything being done with a view to an end, everything is necessarily ordered to the best end. Note well that noses were made to support glasses. That's why we have glasses. The legs are visibly created to be put on. And we have pants. The stones were formed to be carved and with them we can make castles. That is why the monsignor has a very beautiful castle - the greatest baron in the province must be the greatest. And pigs, since they were made to be eaten, we eat pigs all year round. Consequently, those who maintain that everything is fine, have said foolishness. It must be said that everything is the best possible. Candide listened attentively and innocently believed it, for he thought Miss Kunigunda was extremely beautiful,

> even though she never had the audacity to declare herself to her." (*Candide or Optimism*, p. 8)

Obviously, for example, the primary cause of noses being is not the use of glasses and we know it, so there is clear **dramatic irony**.

The lyrical self narrator lets us understand that Candide's desires for Kunigunda keep him in self-selective naivety, which brings us closer to him who has omniscience and also generates, at a certain level, **dramatic irony**.

Due to influences that often corroborate degenerate patterns of action, many take Voltaire as an agent of social change when he produced literary art of very high standards.

When the educated reader is faced with semantic modifications about Leibniz's theories, he does not take the modifications for truth because he knows the truth. The educated reader can be taken by laughter in the face of semantic changes and I believe that this is the effect expected by Voltaire.

Thus, the one who is dissimulating by questioning by pretending not to know the answer is Voltaire, through Candide, and the one who is led to reflect on reality, on his own convictions and to laugh is the one who reads his works through **dramatic irony** that allows the reader's knowledge to be kept adequate to that of the author, which can generate empathy towards the author, which is important to highlight in order to be able to analyze the work *Candide or Optimism*, because it is Voltaire's clear intention to bring new definitions distinct from the common ones about part of the phenomena exposed, definitions that can form caricatures of what is true in order to provoke laughter, to impress or contradict, even if it is by means of sometimes jocular semantic modifications.

The similarities between the works *Euthydemus* and *Candide or Optimism* go beyond the simple use of rhetorical resources. In both, the characters question supposed wise men.

In Plato's dialogue, Socrates, Clinicians and Ctesippus questioned the wise brothers about the science they possessed and, in Voltaire's proposed novel, Candide and other characters question the master Pangloss directly and Leibniz indirectly, to develop a relationship of opposition between the practical life of Candide and the optimism of Leibniz, to live up to the title of the great novel, '*Candide or Optimism*'.

It is important to say that, in Plato's dialogue, Euthydemus and Dionysodorus, they did not have useful knowledge to make the lives of others illustrious, which does not mean that their examples are useless, and they developed a rhetorical technique that could be used to win disputes without having developed excellence, a technique very different from that present in the essays of the philosopher and mathematician Leibniz that are targets of Voltaire.

The definition of irony is very restricted, however, after many studies on the irony phenomenon, it has "ironically" undergone semantic modification by many who have conceptualized it without observing its criteria, although it is true that the phenomenon may have evidenced its existence by its consequences.

When observing a phenomenon a lot, many predicates are commonly generated that restrict it semantically and define it better, however, when observing the rhetorical resource irony, it is noticeable that it has received many semantic excesses to the point that its definitions define other phenomena and other phenomena define what irony is, and when I say this, I do not mean the points of convergence in the zones of semantic intersection between irony and other phenomena.

Not all people have good character and many people avoid contact with the integrality of the phenomenon out of aversion to its consequences, because it is a filtering element of character and revealing truth.

About the result of the study of the phenomenon it is possible to understand that within the ludic there is no deception, just as there is no innate deception in the rhetorical resource irony, thus, unlike those who question pretending to want answers, the deceivers question to bring doubts and their own ideas moved by perversion and not by playfulness to show themselves as light.

The falsifier casts doubts and wrong possibilities as possible before lying or fabricating to pretend to be wise.

When we read Plato's Euthydemus, this is evident in the contradiction of the brothers who only cast doubts in order to create the appropriate rhetorical inadequacy.

So, after acquiring what Plato offers us, it is possible to look at the whole of life in a different way, in a richer way and realize that this tendency to perversion in speech is repeated in all areas of life and it is possible to understand the phenomenon to be able to define and extirpate the types of characteristic tricksters who make themselves bearers of knowledge.

Thus, differentiating what is playful and what is misleading is essential to recognizing the truth about phenomena. Which leads us to the need to observe the phenomena more closely.

III - THE BREADTH OF THE DEFINITION

Irony can contradict and modify semantic charges in a playful way, but both phenomena, contradiction and semantic modification, can be distinct causes and consequences of irony.

What causes irony is, at first, the intention to question in order to contradict what entails contradiction in an "implicit" way and semantic modification in play, because it is true that irony is an element of *play* and not a regular questioning. Even so, many theorists have added to the definition and called irony any contradiction that has a certain degree of implicit elements, playfulness (*play*) and art.

Euthydemus and Dionysodorus implicitly contradicted and caused a playful effect, but the use of the phenomenon of contradiction, with a certain degree of implicit and playful elements, does not allow us to create a formulation of the irony phenomenon without the intention of ironizing. By this, to say, '*So, isn't it?, being a father, is yours, so that it becomes your father, the dog, and you, the brother of the dogs.*' it can be an offense, if there is no playfulness, it can be a statement of truth in a fable, fiction where animals develop speech, and it is irony because of the intention of Dionysodorus who has the will to contradict playfully to cause effect.

Let us note that the referent of the irony phenomenon can be distinct from the words that are used to try to define it and, obviously, many who have theorized about the phenomenon, when creating some new study, have taken the causes that contribute to the existence of irony and the consequences of irony as manifestations of irony or irony, which has led many authors and laymen to call euphemisms, dissimulations and contradictions of irony.

According to many contemporary theorists, an example of popular irony present in everyday life occurs when we are being watched by a camera and there is a message telling us to smile because we are being filmed, when, in fact, it is warning us that we can be identified.

Let us note that this way of speaking, which does not say what one really wants to say and which replaces the heavy burden of warning, has little or nothing of real irony. Perhaps it has only and in part the dissimulatory aspect that is part of the causes that allow the phenomenon to exist, because, as already said, pretending to desire an answer characterizes irony when questioning.

Shall we regard all dissimulations and hypocrisies as irony? Are we ennobling dissimulation by attributing to it the linguistic sign 'irony'?[4]

The semantic load of irony not having very well-defined limits allows offended parties to define the phenomenon in a simple way and in the worst possible way to advocate for their own cause, which can generate, among *amatheís*, undue references to irony.

Since every resource serves mankind, obviously individual intentions must be the source of the irony phenomenon, and when the source undergoes degeneration, [5]the phenomenon that flows from an individual undergoes degeneration as well.

In a state of ignorance and presumption, an individual is unable to define, observe and produce the

---

[4] Typical ironic ancient Greek questioning that serves to defend the truth and leads to rational reflection.

[5] Degeneration is a reference to the loss of one's own qualities. It is commonly said that a concept has degenerated or is vitiated when it has been misdefined.

The Rhetorical Resource Irony

resource and, because he cannot define the rhetorical resource irony, he, when being the victim of some mockery, the possibly hurt person can say the following:

> "Irony has been treated as a figure of refined speech, accessible only to people of great intelligence. However, it is not only a false idea, but even a simple inversion of reality. Irony is the immediate resource of the barbarians; it is the favorite gesture of the ignorant; it is the first impulse of stupidity. Faced with what bothers him, but which he cannot analyze, explain and destroy through reason, the ass laughs. Laughing, he tells himself that the object does not deserve attention, that the interlocutor does not deserve respect, that the subject is not discussed. Then he looks for other donkeys who can laugh with him, because the herd instinct gives him greater psychological security. Laughter is the most practical way for mankind to relieve himself of the feeling of helplessness. We laugh when we see the incomprehensible, the absurd, the bizarre. Any baby knows how to laugh to relieve himself of what would otherwise inspire fear. But only ill-mannered adults make conscious use of this resource to refuse to see the truth."
> (Opinion of a professor of classical studies.), (R. F.)

This opinion is completely adverse to the reality of the phenomenon analyzed in this work and exposes the presumptuous character of its author, so I do not share the same opinion, because I seriously study the phenomenon.

Bad definitions of this type that are limited to judging a resource as good or bad, useful or useless are the result of rudimentary minds and incapable, in the actuality of speech, of developing logical reasoning about some phenomenon. Faced with a phenomenon that they cannot reach, they can only consider the phenomenon as useless for vanity in favor of maintaining their own ignorance and false image of being wise.

It is no coincidence that many of these mediocre definers limit themselves to repeating definitions even though they are unable to understand them. These types become mediocre successors to professors Euthydemus and Dionysodorus.

It is not shameful to know little and define little, in fact that the little definition exposed is the result of knowing and knowing really is to live the association between the definition and its referent, the linguistic signs and their referential phenomena.

It is remarkable that, among the wise, irony allowed to act against offenses that were the result of ignorance masquerading as erudition, and laughing to advocate one's own cause was never defined, in fact, as irony at any time, although laughing can be a consequence of irony, but never as a resource that is used only among the proud ignorant, As has been proven, to define the rhetorical resource irony in a way that is appropriate to a defense in one's own cause and to observe only the bad examples of the bad ones to define a phenomenon is dishonest action.

The Rhetorical Resource Irony

Esteem for interlocutors can be an unperceived phenomenon, however, to develop a disrespectful attitude in this way it is necessary to conceive the interlocutors within a lower common standard. Many use their own measurements to measure the knowledge and attitudes of others or speculate to depreciate because they consider that there is also pretense and mediocrity among their opponents. This is a reflection of personal choices that demonstrate who people really are when they define phenomena for themselves.

Euthydemus and Dionysodorus created rhetorical inadequacies common among degenerates, who make use of them for self-defense or self-promotion, and their attitude does not even represent the completeness of their acts, much less the definition of the phenomenon studied.

Linguistic signs with their modified semantic loads, obviously, generate more phenomena of semantic change and the use of semantic modifications to elevate vice to the level of rule is a very common vice among degenerates who advocate their own cause in their productions and this mania does not characterize irony or at least should not characterize it.

The linguistic sign 'love' that received the semantic charge of the word 'greed' and the word 'idolatry' exemplifies the phenomenon of vice degenerating what is noble, that is, attributing to the linguistic sign 'love' the semantic load of what the being conceives as love that can be the greed of one's own heart. This is an extraordinarily verifiable truth, for among the degenerates it is said that love is felt for the targets of their lusts, which shows that semantic modifications of this kind are not to be confused with playful manifestations, because they are distinct phenomena.

Among the educated, it may be playful to call greed love, but among the degenerate, greed is love and there is no playfulness in semantic change. This phenomenon is very common and with the linguistic sign 'irony' something similar occurred.

For Muecke, irony has the function of balancing and correcting the level of seriousness of life. *"Stabilizing the unstable, but also destabilizing the overly stable."* (Muecke 1995, p. 19)

Not exactly in this way, because we naturally ironize to destabilize when we do not have respect, however, I agree that the irony phenomenon can serve to remove seriousness, in the case of Dionysodorus' irony that serves for playfulness. Differently and for other reasons, sarcastic irony produces real offenses, because, as already seen, offenses occur in some way, even if they are not verbalized, so when it does not serve to evidence the truth, sarcasm generates a false sick humor, because it works against the truth.

I do not agree that irony serves to promote a balance, even though I am aware that the motives that move individuals are particular. Obviously, Socrates' primary intention was not to destabilize the brothers' speech, it was to highlight the truth about their speech by provoking them to offer them what he wanted, which could generate the desired effect, obtain an eristic sample and, consequently, destabilize the deceptive discourse that is a consequence of evidencing the truth. however, destabilizing the deceptions or promoting a friendlier situation that allows playfulness is not a primary consequence of irony and neither is the intention of Socrates in the specific case, although he made his own laugh by bringing relief to the tense atmosphere, however relief is not the reason for irony and neither is what Euthydemus and Dionysodorus felt, and this is provable, because Socrates ironically admonishes his people when he asks them why they are laughing at such serious and beautiful things.

> "And Ctesippus, as was his custom, laughing out loud [...] And Clinias, all delighted, began to laugh, so that Ctesippus became more than ten times greater.
> I (Socrates) said: Why do you laugh, Clinias, at such serious and beautiful things?" (Euthydemus, p. 117, 119)

Clinias laughed even more at Socrates' questioning: "Why do you laugh, Clinias, at such serious and beautiful things? (Evidence of playful and ugly things.)".

Knowing that the brothers' questions were not serious or beautiful, Socrates possibly sarcastically questioned why they laughed so that everyone would laugh even more.

This is a completely playful way of questioning coming from Socrates by pretending to want an answer. Let us note that Socrates does not create rhetorical inadequacies even when developing sarcastic irony, unlike his brothers.

When advocating in their own cause, deceivers do not conceive the truth as something integral, so they do not consider a moral basis when expressing themselves and, consequently, because they do not accept a single truth, destabilization becomes the primary motive that advocates in favor of their own character or the idea defended.

In this case, in a dispute without a base, the winner is the one who plays or deceives better and not the one who is right.

If Euthydemus and Dionysodorus were to quarrel, it would not be the one who is wiser, but the one who best elaborates the deception, because the deceivers full of presumption have their own techniques and definitions to form particular certainties.

In dispute, the true search is not the balance of ideas or a humorous tone to balance a serious dispute, nor does humor characterize irony to take away the seriousness of a dispute. Socrates used sarcastic irony when it was no longer possible to speak seriously.

If we look for the intentions of the participants in the dispute, we will have different reasons. The brothers sought the reputation of being able to teach virtue, excellence; Socrates sought a sample of eristics from the two; Clinias, being inexperienced, sought to learn something and develop his own argumentative technique and Ctesippus sought to learn and possibly grow on his brothers.

Being caught in contradiction in serious discourse greatly tarnishes one's own image, so which of these used the rhetorical resource irony to lighten the heavy load of the dispute in a primary way? Only Dionysodorus, because he had a problem in his hands given by Socrates, because he wanted to be in a playful environment when he was caught in a possible loss of reason, because what is playful works in a time and place outside of reality, which gives it the quality of a place of escape.

It may be that there are, among ideologues, those who do not believe that there is a truth, those who feel uncomfortable with the harsh reality and who want balance for the sake of balance because they do not accept the truth, but this insane posture is very particular to ideologue groups and individuals who generally defend what they supposedly prefer and not the truth.

Voltaire was right who evidenced cowardice in the philosophers of his time.

In practice, the idealized balance is not as beautiful or romantic as it is presented in theory, because ideologies act as an agent that removes sensibility.

If irony serves to contradict for the sake of contradicting, if it aims to promote a balance for the sake of balance, it can disregard the importance of the existence of a truth to be defended and the importance of part of the personal motivations for balance, which would be absurd to say, because it would remove the real reason for its use, including its playful motive.

In the initial time of a dispute, there is no irony for balance or relief in serious speech. Dionysodorus, by not using serious speech, created inadequacies with misleading questions at the beginning of the disputes against Clinias and Ctesippus, which is a particular characteristic of the deceivers and not of the phenomenon exposed.

The reason languages are created is to communicate reality, something that also exists in different ways when we promote playfulness, so each definition, playful or not, needs definition or ignorance about space, time, forces and masses involved implicitly or explicitly.

Are many scholars of the phenomenon removing the playful charge of Greek irony by taking sarcasm as not useful for the evidence of the truth and contradicting for the sake of contradicting for the promotion of a balance or relief in the spirits as resources for the evidence of the truth? I think so.

Voltaire used the rhetorical resource of irony several times in his journalistic and literary productions, when he developed writings that showed that there are traitors and degenerates in any institution regardless of its rules. Would this promote a balance, truth through discord or laughter? All three elements are possible.

Judging an institution for its traitors, which Voltaire did not do, is defamation and would provoke hatred from people who did not understand the *game* developed by him and would not provoke a balance between the representatives of the offended institution.

If we consider that the balance is in the idea of what is addressed, the intention would also be to defame and the effect is the same as already evidenced.

This shows that the balance referred to only works between disputants exactly in their moods, without being useful to evidence the truth, because what is playful works *in a proper time and place, which is neither ordinary life nor real life*, according to Johan Huizinga, which gives it the use of escape.

Voltaire simply created possible, impossible and inconvenient fictional jocular situations for the promotion of laughter that led him to suffer various adversities. Voltaire was never a revolutionary who hated it and promoting laughter semantically is next to charity.

There is a useful aspect to mockery which is the refinement of the good, but many observations make me not consider this motif very prominent in some of Voltaire's writings. I think that, in his great novel, as in many of his publications, Voltaire used the resource of irony for the sake of satirizing and not for deception or for the disclosure of the truth. I am sure that *Candide or Optimism* is a novel written to provoke laughter by satirizing the society and values of its time and not to expose truths, although there are truths exposed about phenomena present in it.

Candide or Optimism should be appreciated as literary art and not as information literature, however, obviously, when reading fictions it is possible to assimilate some value from them, which requires understanding, knowing how to identify and use the rhetorical resources that provide complete understanding.

Voltaire worked in his writings several human values and many complex concepts such as Leibniz's theory of the best of all possible worlds in a simple way and, to this end, criticized the complexity of the theory as if it were confusion, creating a "counterbalance" that, according to Muecke, would have the objective of achieving a balance.

According to Johannes Huizinga in his work *Homo Ludens*, the ludic serves to relieve tensions and works in a place with its own time that is neither ordinary life nor real life, which contradicts any definition of something truly ludic to balance reality, to promote something that is not mental relief.

When thinking with Muecke, Voltaire would be a promoter of social change, however I do not find this reason in reality, because he does not have the bad education to bring elements of fiction to reality. Realizing the importance of a "balance" of ideas or aiming is in itself a misleading perception, the truth being only one, and I do not believe that words can express reality completely.

Relief is important because it timely and appropriately undoes the seriousness of matters, yet they have no bearing on true definitions or conclusions.

We must realize the importance of what is playful for the relief of individual tensions and not as something shared as a balance of ideas.

What is facetious is not dealt with as a source of information. What faithful information could be drawn from something that was made to differ from what is real as what is playful? Just repetitions of what is known as what provides the sarcastic irony and thus draw the possible conclusions as a byproduct of interpretation. Obviously I am referring to the phenomenon and I know that, outside the ludic universe or in the transition between it and reality, many values can be assimilated.

Let's see what Pococurante (Voltaire's character) says about the radical socialist revolutionary ideological evil that prevents people from speaking the truth because they follow idealisms:

> "Yes," replied Pococurante, "it is beautiful to write what one thinks; it is the privilege of man. Throughout our Italy, only what is not thought of is written; those who inhabit the homeland of the Caesars and the Antonines do not seek to have an idea without the permission of a Jacobin. I would be very glad of the liberty which English geniuses inspire if passion and factious spirit (partisan passion) did not corrupt all that is estimable of this precious liberty."

The Rhetorical Resource Irony

In this case, as in others, reality is so absurd that to talk about it truthfully provokes laughter, Voltaire's primary intention in his great novel. The reason for creating playful situations is to provoke laughter, with the most unusual situations being those that undo concepts and values more ostentatiously in play. They are the most impressive and, consequently, the ones that can be funniest the most, which can generate definition and not true criticism such as the truth exposed about the Jacobin reality that does not allow freedom and idealizes promoting freedom.

The balance that comes from a lie well written as truth always has a hidden motive because it must have its origin in the immoral and the playful does not support this, even the character of the text is at the mercy of the character and arbitrariness of its author.

Literature tends to have many mistakes that receive the refinement of beautiful and strong words. It is in literature or in the arts in general where deceptions can be confused as part of reality, because, by mixing fictional elements with real ones, it is possible to bring greater verisimilitude to fictional elements that can influence current and real life, causing change in the individual conceptions of the unwary who are led to deception, when they should be entertained.

True literary art serves to promote beauty, because the reason for art to exist is to spread what is worthy and beautiful and, as has already been said, through the rhetorical resource of irony, the depreciation of something beautiful can further highlight its beauty and value.

In literature, the contradictions of the situations narrated have also been called 'irony'. When a character's thought or speech contradicts a situation, even without interaction with another character, there is semantic amplitude, because the contradiction is perceived by the reader. Thus, **situational irony** could occur between the text and the reader.

According to Muecke (1995), literary irony can be divided into two major categories, **situational or observable** irony and **verbal or** instrumental **irony**.

According to the theorists of the phenomenon, the speech of a character that contradicts an expressed, observable situation, I would say that it is also presupposed or entailed, is **situational irony**, so we are taking as irony a contradiction in what is narrated and this way of perceiving the phenomenon expands the definition, creates an addition that, theoretically, works in favor of its evidence.

Lines that are contradicted by the situation occur in a significant way in the novel *Candide or Optimism* and the reason they exist is the promotion of laughter, a common consequence of ancient Greek irony.

The criteria used to consider additions as irony are simple to define. It has been observed that this type of speech contradicts and can persuade, cause astonishment or provoke laughter, however the phenomenon is very complex and it is not possible to formulate it without a context. At some point, after reading a statement that contradicts a situation, there is the formation of questioning about what was said to generate surprise and change of mind that is not necessarily contradictory, which characterizes the existence of the ironic in these cases.

For some of the theorists of the phenomenon, there is also the literary irony called **instrumental** or **verbal** that occurs when there is what they call semantic inversion, that is, a change in meaning. It occurs when the author changes the semantic load completely or partially from icon or linguistic sign to the primary intentions of persuading, mocking, or provoking laughter.

It is notable that they are considering as irony elements that work for the phenomenon to exist, because semantic modification is necessary for the irony that provokes laughter to exist.

Let us note that the focus of understanding is on the inversion of the semantic load of linguistic signs and not on the reason for the inversion or on what the inversions provide, which can be a series of phenomena that can include developing irony, which shows that theorists are predicating playful intentional semantic modifications as irony, in fact that it is necessary to have the intention of ironizing for something to be ironic and semantic modification would be a resource that promotes irony the intention.

When a character says that something is perfect when it is perceptible to an interlocutor or reader that it is not finished, there is what is considered **situational irony,** because the situation described is not the one observable to an interlocutor or reader and there is **instrumental irony** in the use of the word 'perfect', when the observable is incomplete and, If the interlocutors consider that perfection exists and imperfection is perceived only by the reader who does not participate, there is **dramatic irony**.

In this case, irony only occurs if the linguistic sign 'imperfect' is replaced by 'perfect' by the one who affirms, in the case of **situational irony**, and naively, in the case of **dramatic irony**.

It is observable that **verbal irony** and **situational irony** complement each other and this is due to the fact that many studies have been developed that have observed the causes and consequences of the irony phenomenon as a rhetorical resource, which has led to the call series of verbal constructions that promote irony, along with the possible consequences of irony, irony or ironic.

There is no mistake in saying that something is ironic when it contributes to the occurrence of irony in some way.

Let us note that what is currently considered **instrumental irony** is a necessary semantic modification that contributes to the joking modification of the meaning, that is, what is called **instrumental irony** is an element that can contribute to irony and not one's own, because irony has the reason for the existence of the conduct of other people's thoughts and not all semantic modifications are the result of the rhetorical resource of irony.

Irony is a resource that transcends from intention to a verbalized form and what we verbalize could not, if it had isolated and decontextualized parts, evidence the phenomenon of irony, so the evidence of the phenomenon is always contextualized and difficult to define.

Obviously, modifying the semantic load of an expression can contribute to creating the irony phenomenon, because modifications can persuade, but we must always be aware that the irony phenomenon is not a set of written words and that written or spoken words serve to refer to it or to invoke it.

Regarding that semantic modifications and speeches that contradict situations can contribute to the existence of the resource in a literary work, I say that semantic modifications can be ironic, as has already been evidenced when I spoke about the inadequacies necessary for the phenomenon to exist.

When we read a novel and the narrator informs us that what lies ahead is an indescribable horror and the reader does not feel the horror announced by his technique not having been good enough to do so, there is no irony. It happens that, if, in the same situation, there is the writer's intention to create an expectation that is not fulfilled to promote some unexpected situation or jocularity, there is ironic regardless of the opinion or effect caused in the reader, which shows that intention is essential for what is ironic to exist. We have the use of the same words in different definitions, as ironic or not, depending on the author's intention.

In this case, there is no guarantee that the effect of contradicting the idea of the reader or of any interlocutor will be produced, and this shows that the current semantic breadth may not be applied in certain cases, when we have only the text as a basis and also shows that there is a need for historical contextualization when we evidence the phenomenon in a corpus of narratives.

When considering the breadth of the definition, it is possible to say that, in his great novel, Voltaire made use of **instrumental irony** to modify the complex definition of 'optimum' evidenced by Leibniz, by providing the expression 'best possible' with a semantic charge that does not allow a world with afflictions without a superior material good to bring, in the reality of the narrative, playfulness to Leibniz's theory that can affect the characters and the reader during the pact of verisimilitude, causing laughter.

The modified reason of the characters ignores the past evils for future equivalent or superior lesser goods and, obviously, those who have not read the work *Essays on Theodicy* will not detect the semantic change in the definition of 'optimum' and, consequently, may consider that Voltaire evidenced a derogatory truth about some point of Leibniz's theory exposed.

Certainly this is not the effect desired by Voltaire. Because they are inadequate semantic modifications in fictional text, the new definitions should not be used in the ideas of current and real life without criteria, as an unquestionable source of truth.

Knowing this makes it possible to understand that modifying the semantics of words is only a resource that contributes to the existence of irony and its critical form, satire, and that semantic modifications can contribute to other phenomena, which shows that it is necessary much more than modifying the semantic loads of words to make use of irony.

In a playful text, phenomena that have undergone real semantic modifications can be mixed with fictional ones, however, the deception is not to be confused with the ironic and is simple to detect, because when what is said is not easily perceptibly opposed, possibly the rhetoric contributes to the deception.

In irony, the modifications in the semantic loads are generally more evident, because in order to produce both the questioning of ideas and laughter, it is necessary to be aware of the two poles of meaning involved, the real possible and the playful, which increases the possibility of the interlocutors understanding the phenomenon in question with a minimally necessary amplitude.

In mistake, there is usually a partial inversion that is difficult to detect, because it has the intention of modifying, of forming a new reality with some level of verisimilitude with the real, which differs it from the ludic that always happens in its own time and place outside our reality.

Instrumental **irony** or semantic modification often occurs in the novel *Candide or Optimism* in a partial way, because Voltaire changes the semantic load from the expression 'best of all possible worlds' to another that does not allow the afflictions of the world without a superior material compensation, which would lead to the conclusion that Leibniz's theory is flawed if the reader made the mistake of confusing fictional elements with real ones because he had not read it *Essays on Theodicy,* or because it considers that it is possible to understand or have access to all the compensatory phenomena of evils.

Voltaire does not seem to invert the meanings to the point of being detected by laymen as those modified to provoke laughter. On the other hand, there is a possibility that he wrote to the few who could detect it, to educated people.

An example of the characters accepting the new definition occurs when Candide, still deluded and agreeing with his master Pangloss, states that everything is fine. How can you see:

> "You are right," said Candide: "That is what Pangloss has always told me, and in fact I see that everything is in the best possible way." (*Candide or Optimism*, p. 11)

In this fragment, Cândido concludes that he ran the best possible way because something good happened to him after suffering, when, in Leibniz's theory, the evil of the world is something that is part of it and the improvement does not take place in the individual sphere in such a way that it can always be perceived or as the characters conceive, that is, The balance of good and evil is for the world and not for a consciousness disassociated from the idea of the completeness of the world.

In the fragment, it is noted the rhetorical resource to be maintained through shallow analyses that do not touch the real phenomenon studied by Leibniz in his essays, analyses that lead part of the characters and readers to repeat fallacies and be deceived, just as the young and naïve Candide was deceived, which guarantees playfulness and satire when we read the work, because many naïve or not, will accept, when reading, the definition accepted by the naïve Candide.

It is true that Voltaire did not bring the truth about Leibniz's theory and did so to put it in a playful situation, in satire, however, not bringing the truth about a phenomenon to develop laughter or *play* against it does not imply that the definition of the phenomenon is or is not assertive.

Voltaire laughed at the philosophers who recruited and those who were recruited in his time. The satirist is to ridicule by saying, implicitly, that the one who teaches misrepresents or does not know what he is talking about and what he hears and approves of what the master says is naïve, deceived into laughing at the blindness of recognition.

The modification that contributes to the hoax is always semantically incomplete, because it is made to resemble a truth, however, the ironic is also ironic, so the contextualized intention must provide what is necessary to detect the reality of the phenomenon.

Irony must be possible to detect and what guarantees its evidence is logical reasoning and the correct foundation, phenomena that are not owned by most people at the same level.

Ill-mannered people interpreting jocular texts with a high level of irony repeat elements of Voltaire's fictions as if they were indisputable truths.

In this reality, where everything is doomed to perish, would it be possible to maintain a fixed idea about something without taking into account its degradation?

In the observance of phenomena and characters, it is observed that there are samples of different manifestations in different contexts and the facetious is often to observe the worst aspect and use it to represent the whole in all contexts, which is also the property of people who form the majority devoid of capacity and good character to analyze, however, using the worst example to represent the whole would contribute to satirizing more effectively in favor of laughter among the more educated, and Voltaire understood this. Voltaire clearly knew that, for example, joining two predicates or adjuncts of nouns, one common or desirable in initial position and a second derogatory and/or contradictory apparent graphically or audible or entailed or presupposed, in addition to promoting rizo can contribute with an idea or convince, which is ironic and used in definitions of many of his characters.

For all that has been exposed, I think it is possible to understand what irony is and what is necessary for it to occur, which contributes to understanding and analyzing the proposed works preceded by some historical basis and contextualization, by making use of the Aristotelian method whenever necessary to be able to confirm statements and the use of the rhetorical resource irony on them.

As has been demonstrated, theoretical bases are necessary to promote a complete and assertive critical analysis and, by having exposed the necessary theoretical bases, it is necessary to characterize the genres exposed and historical contextualization of the ironic manifestations to be analyzed in order to be able to understand the intentions of the ironic productions and part of their values.

IV – CRITICAL ANALYSIS

Characteristics of the Literary Genres Short Story and Novel

When defining the Greek rhetorical resource of the Ancient Age, a Greek work from the Ancient Age was used to adapt the linguistic sign 'irony' to its referent and to be aware of the correct semantic load of the phenomenon. To analyze the resource in works of the Modern Age, it is necessary to expose some characteristics of the textual genres used and contextualize the reading historically. It is necessary to highlight some aspects of the literary genres short story and novel so that there is a better understanding of the works that will be analyzed.

Theorists of the short story literary genre always compare the short story genre with the novel to create distinctions and I consider this fair. The modern novel is a fictional narrative written in prose that has facts created by or related to reality. In it, everything can be detailed, as there is nothing to prevent the author from describing his fictional elements unnecessarily, so that they do not contribute to the plot, many of them having only an aesthetic character. Thus, it is possible to find in the novel: long descriptions of characteristics, excerpts of songs to bring charm or inform the status of some phenomenon, personal tastes of the characters, values that do not add to the plot or conclusions, and many other elements that can be merely illustrative as several protagonists that do not conclude any idea.

Unlike the novel, the short story does not allow merely illustrative items or they are less common, because there is a physical limitation, that is, in the short story, everything must contribute to the narrative, because, otherwise, we would have some disconnected passage within another small text that would be counterproductive and perhaps unpleasant, if this is not its proposal.

Due to the lack of physical limit, the novel genre allows the development of greater literary richness and, consequently, better development of the **rhetorical resource irony** and this fact will be evidenced later with the exhibition of a great irony maintained by several ironic elements that contribute with a main irony and elements fruit of it.

According to Cortázar (1974), a good short story must be incisive, mordant and intense. It must manifest itself in this way from the first words or scenes and must not contain gratuitous elements that are merely decorative due to physical limitation.

The first characteristic of the short story is its physical limit, which is something around thirty pages. Above that, fictional narrative is categorized as a novel (long story) or novel (text in proza).

Obviously and ironically, the simplest definition, due to physical limitations, tends to agglutinate more definitions and possibilities than the more complex definition that limits.

For Cortázar (1974), in the short story, the idea of signification cannot make sense if it is not related to that of intensity and tension, that is, when we manage to reconcile meaning, tension and meaning, we obtain a true tale, *agglutinating a reality infinitely vaster than that of his mere argument* that makes us realize that there is much more than the apparent brevity of the physical limit, which allows for the existence of many layers of meaning.

The Rhetorical Resource Irony

Being incisive and mordant is a need that serves to adapt the text to the format that, despite being limited, cannot be perceived as limiting the amount of meaning. Art in writing is about giving depth to the text that makes us conceive of the narrated universe as something with many possibilities, so that it is possible to lose even the notion of the physical size of the text.

A short story can provide more memorable and striking elements than a novel that is often longer.

When the author is writing a short story, he may realize that he can generate a novel from it by extending it, or he can make use of elements of a novel to write a short story. This aspect will be evidenced by analyzing the proposed works, the novel and the short stories, which have certain shared elements.

A striking feature of the tale is its apex or ending. In a novel, most of the values and conclusions can be worked out during its extension, so that it is possible to have several peaks beyond what exists at the end, which does not normally occur in the short story, because, due to its physical limit, it usually has its apex at the end or close to it. Some authors use time games and reorganize their apexes in both genres depending on the literary art.

Novels and short stories can share characteristics of other genres in a moderate way and everything that is said about the novel and the short story can vary from author to author, being the most traditional maintainers of the quality, value and beauty of literary art.

The less traditional, faced with the inability to produce art, produce works with the intention of shocking and breaking moral taboos, however, works with modern, non-traditional proposals may be literary art in some way and works with traditional proposals may not have quality, value or beauty, because the quality depends on the author's ability.

Regarding the transfer of elements from fiction to reality, both the novel and the short story are propitious of the phenomenon, but it is the novel much more propitiating details that can lead to the encounter of great complexity, as will be evidenced in the novel *Candide or Optimism*.

It may be remarkable to some that, in the novel and in the proposed short stories, Voltaire's search was not like that of the great philosophers who sought truth and virtue, that his motives were to promote playfulness, to laugh, to contradict and only to entertain, however, obviously, these evidenceable characteristics are restricted to the universe of each work and there is the cultivation of what is beautiful in his proposed works and in many others and this is something that It must be recognized, because values are manifested in some way in artistic productions and never could a work accommodate them all with a restricted proposal.

For some without the proper education, its semantic modifications present in manifestations of literary art have added possibilities that may be unexpected and degrading to the truth, serving to shock and break moral taboos, but this is a common mistake.

Will they prohibit narrating traitors, murderers and suicides because of the lie that the writer is inducing such postures?

Obviously, after reading, after the pact of verisimilitude is broken, educated people return to ordinary and real life without carrying elements with modified semantics from the fictional to carrying only the values that they deem assertive by their will.

The Rhetorical Resource Irony

Historical Contextualization - The Illuminist Voltaire[6]

For Aristotle, who exemplified through his work what a scientific essay is, in an analysis there must be an analytical category about a *corpus* that is examined in the light of an immutable truth under a given circumstance. This set should be used as many times as necessary to overcome existing doubts and remove all possibilities that do not agree with the possible truth referenced or not, which I consider sufficient to analyze any phenomenon. From this perspective, science, regardless of its evolution, provides the basis and not the conclusion.

The focus on the analytical category should never be diverted, because the deviation would lead to a lack of meaning for a perfect, complete and assertive reading, although it is appropriate to use analytical categories that precede and are involved with the analyzed phenomenon, the rhetorical resource irony, to contribute to a more complete view.

Despite all the existing techniques, when there is no truth in the speech, that is, when there is no adequacy, any scientific essay can show deception as truth.

As has been said, all literary analysis must have a basis and, through this writing analyzing some works of Voltaire, I realized the need to previously expose some data that contextualize the corpus historically beyond the exposed phenomenon.

---

[6] The following text has shared parts of *VERUS DEUS*, Introduction to Christianity, 2016 and *MAXIMO*, 2020 and has been added as a complement in recent editions.

There is no complete reading or close to being a complete reading without historical contextualization and the case of Voltaire, the admittedly naïve who was enticed as a young man is interesting and elucidates the Enlightenment influences on him, so that it allows us to analyze his works more thoroughly.

François-Marie Arouet was born on November 21, 1694, in Paris, France, to parents who were members of the French aristocracy. He lived in modernity and enjoyed the intellectual growth cultivated in the previous period, the Middle Ages, which required, on the basis of his studies, a strong learning of grammar, rhetoric and logic, which explains his great mastery of rhetoric.

He studied at the Jesuit College Louis le Grand where he had a classical education. He learned dialectics and theology, tasted a great didactic in the mold of Catholic universities, but had his customs spoiled at a young age by bad influences.

Due to his new inclinations, Voltaire was lured at the age of twelve and became a member of the Society of Temple Libertines where he became addicted to the unruly pleasures of life, in the year 1706, a cause that made him develop a temporary aversion to morality.

His father and brother, both very Catholic, were afraid that the naïve François had spoiled the customs, so they arranged for him to move to the city of Caen, France, in order to distance him from the bad local influences. In Caen, he worked in an office.

His joviality passed and with it also his naivety that allowed him to be enticed by the Enlightenment philosophers.

In his maturation, he took pleasure in expressing the idea that everything had an opposite side

The Rhetorical Resource Irony

in itself, that is, he had a tendency to contradict and satirize.

It was then that he realized the need for irony to respond with mockery to what does not deserve respect, despite also what deserves respect.

In his productions, he liked to contradict formed concepts and created real, possible and fictitious ironies, so in 1716 he was exiled to Sukky-sur-Loire, after writing two epigrams about the regent Duke of Orleans whose content contradicted his morals and personal reason, characteristic of satire. The following year, 1717, he was arrested for satirizing the regent of France, imprisoned for eleven months and began to write the poem Heuriad and finished Oedipus.

In 1718, François named himself Voltaire, the same year that his play Oedipus earned him great success and he inherited a large fortune.

From that point on, the lies against him intensified and, according to his words, there was a veritable library of lies against him where they attributed to him lines and texts.

Let's look at François-Marie Arouet in self-defense against the slanders of an ungrateful traitor to the Catholic Church who had received help from him as a brother in the faith, of a heretical theologian who misrepresented his writings and speeches and disseminated statements about him not based on facts, and of one who found a market niche to produce and sell slanders against him:

> "If I dared to include myself among those whose works only obtained persecution as a reward, I would make me see a band of wretches obsessed with ruining me

> from the day I presented the tragedy of Oedipus; a library of ridiculous slanders published against me: a former Jesuit, whom I had saved from the last torture, by rewarding me for my service with defamatory libels; a man, even more guilty, printing my work on The Century of Louis XIV with notes in which the crassest ignorance spews out the most infamous impostures; another who sells to a bookseller chapters of an alleged universal history attributed to me, and who, eager to print such a shapeless fabric of nonsense, false dates, facts and crippled names; in short, cowardly and perverse enough men to impute to me the publication of all this rhapsody. It would make you see society infected by this kind of men, unknown to all antiquity, and who, not being able to embrace an honest profession, either as a commandant or as a lackey, but unfortunately knowing how to read and write, set themselves up as spokesmen for literature, live off our works, steal manuscripts, deface them, and sell them. (Voltaire, to Mr. J. J. Rousseau, in Paris, Les Délices, August 30, 1755)

The traitorous Jesuit cited by Voltaire is Pierre-François Guyot Desfontaines (1685-1745), a journalist and critic accused of sodomy in 1724. He treacherously paid for the help that Voltaire gave him to get out of prison. 'The man, even more guilty' is a reference to Maupertuis, another heretic creator of theology who kept saying that Voltaire spoke ill of people to authorities such as the king of France.

The Rhetorical Resource Irony

In 1726, he was arrested and then exiled after quarreling with the nobleman Guy-Auguste de Rohan-Chabot.

Here is a literary legend without historical proof that I heard from my literature teacher Arturo Gouveia. Arturo gave Voltaire an additional imprisonment for telling a woman of the nobility that he wanted to see a noblewoman menstruate blue in proof of nobility. I remember laughing a lot at these innocent jokes. The problem lay in religious Pharisaism. He heard from brothers more aligned with religious fanaticism various lies against Voltaire as one of his favorite scarecrows.

When they did not create texts and attribute them to him, the traitors misrepresented his writings with footnotes and interpretations to be used against the Church. During this writing they repeat lies against Voltaire without knowing that they repeat lies of degenerate pharisaical liars at best and that they have empathy for them. To this day it is said that Voltaire, when signing his productions, complemented his signature with an expression against the Catholic Church.

Today, Protestants[7] have a veritable library of lies against Voltaire and many members of public life and lies always lead to persecution, so they must be evidenced.

Voltaire had to ask forgiveness even for sins he did not commit and today many illiterate people, who read his works to confuse fictional elements in them with reality, blame him.

---

[7] 'Protestants', in this case, is anyone who protests against Voltaire, it is not a reference to those who protested against the Catholic Church.

Years later he was still provoking anger in various representatives of public life. In 1729, he returned to France and wrote several tragedies inspired by Shakespeare's work. And they are: *Brutes* (1730), *Zaira* (1732), *Adelaide du Gueslin* (1734), *The History of Charles XII and The Temple of Good Taste* (1733). In London, the French version of his work *Philosophical Letters was* seized and burned, and he took refuge in the castle of Cirey de Mme du Châteler, his great passion that lasted sixteen years. In 1754, Voltaire wrote many plays to be staged in the theatre of Cirey Castle. It is concluded that if there were no contradictory elements in public life that served as an example to him, Voltaire tried to create them when he produced. In his works, he invented many fictional characters to contradict concepts formed about phenomena, so that some marvel to this day at his creativity that he managed to bring from fiction to popular knowledge, as reality, some elements of his creation.

In 1746, Voltaire became court poet and, due to the prestige acquired with the position, he became historiographer of King Louis XV. In 1747, Voltaire committed several serious imprudences, a combination of factors: satire and indiscreet commentary, conflicts with literary rivals, excessive political ambition and involvement in unofficial diplomatic negotiations.

Voltaire had a habit of believing that he could manipulate the court with pure intelligence while the French court functioned as a swamp of armed vanities.

At this time he got too close to the royal circle, gained violent envy, participated in frowned upon political maneuvers and ended up becoming vulnerable to attacks and suspicions, so he took refuge for some time and went into melancholy.

There is also the episode linked to the book *La Princesse de Navarre* and life at the court of Versailles, where he sought official recognition but encountered constant humiliation and rivalry.

The "refuge" and the "melancholy" are probably due to the psychological wear and tear of this period, a sense of persecution, frustration with the court, nervous and physical crises, and disillusionment with the aristocratic environment.

Voltaire then goes on to develop a much deeper resentment.

Meanwhile, the Pharisaics invented lies against him out of envy, including that he was a thief for having bought tickets in batches, because he understood arithmetic, which guaranteed him profit.

Because of new publications in this period, Voltaire was arrested in Frankfurt and then wrote *Essays on Morals*. After turning sixty, he set up a theater where he exhibited several plays around 1755.

In 1759, Voltaire broke with Rousseau, who was an Enlightenment philosopher, therefore a brother in concord, and wrote his great work *Candide or Optimism* ironizing everything he could to promote derision and debauchery against the philosophers of his time. A novel that serves as an example of his ironic way of writing and as revenge to dissolve his resentment against everything that acted against him.

He regretted the loss of friendship, however, he had another path for a long time.

At the age of eighty-four, Voltaire returns to Paris, is feted and attends a performance of his tragedy Irene at the Academy. He had the pleasure of having his work recognized during his lifetime, and before he died he wrote the following: **"I die worshipping God,**

loving my friends, not hating my enemies, detesting superstition."

Armed with all the memories of those who had acted against him, he had not abandoned faith in God, he maintained respect for God and laughed at the priests who dishonored the faith, the philosophers who did not live the ideals they pronounced, the postures and pretended positions of men.

Like every Illuminist, as long as he was associated with the degenerates, he loved the supposed reason that was probably the set of his convictions as naïve, inexperienced, enticed and deceived by libertine philosophers, which gave him good material for his ironies about cowardly philosophers incapable of governing their own existence and who, despite this, felt able to advise. Fortunately, Voltaire really woke up from the philosophical delirium of considering enlightenment to be idle for ideas.

For many, Voltaire was a great promoter of social change and he did so by presenting unusual perspectives in his productions, which is a big lie. In fact, he did not change anything in the government or society of his time. He changed many individual hearts by planting a seed of sincerity, freedom and, consequently, love.

Certainly Voltaire contributed to the discrediting and devaluing of the targets of his criticism. On the educated, those capable of detecting his ironies, he infused no evil effect, which shows that evil accompanies ignorant people in a superior way.

Voltaire's temporary desire to break with the knowledge inherited by the Catholic Church in the medieval period categorized him as an Illuminist, however, he soon abandoned vagrancy and returned to his father's sweet lap.

In their writings, Protestants were all placed in a bestialized milieu where Christ is quoted only for the love of discord and self-interest, an environ where one execrates and kills oneself for differences.

The lyrics created by Voltaire show a common root between Protestantism, the Enlightenment, and the notion of a particular god whose concept is marked by human arbitrariness.

Voltaire laughed, mocked everything he could and offended a lot. He had a greater effort to mock philosophy, Christianity and morally high ideals, he produced many philosophers alienated from real life and priests who were not truly saints. All scoundrels, corrupt, opportunists, swindlers, fools, egocentric and especially promiscuous. He did this perhaps because there was no interest in what can be called sameness or out of pure jocularity that affected even those who honor ecclesial garments.

Voltaire created characters who were members of the Church who were adverse to their own convictions, however, one cannot judge the character of a man by the lyrical selves he produces, by his literary art.

He was already in the arms of the Holy Catholic Church when he fell ill and, after vomiting blood, he tried to confess to a priest who came to him.

After communion, he obtained an improvement and continued to produce his literary art until the end of his days. He died faithful to Catholicism and sanctified by God.

Voltaire maintained the entirety of his works, which include *Candide or Optimism,* and did not recognize them as offenses against the holy church, which shows that the purpose of the aforementioned novel, as well as other productions of his, was not to tarnish the holy church.

We must bear in mind that the historical facts alive in popular knowledge of Voltaire's time mainly involved philosophy, Christianity and their educational influences, so he would have to satirize these elements in some way by his art.

He would mainly satirize philosophy and Christianity for his past engagement in the new Enlightenment philosophical awakening, which possibly provided him with the model of a cowardly philosopher alienated from ordinary and real life.

Philosophy among the Enlightenment was somewhat ineffective in the search for truth, so in his great novel, Voltaire defined the character of the philosopher as that of the coward incapable of managing his own life who perceives himself able to advise others, their action against the villainy of the world.

There were two distinct paths in philosophy, just as there are today. One of them sought truth and virtue and the other fame, winning for the sake of winning. The opposition between these paths was very clear, although the pursuit of virtue can lead to fame, but if the goal is fame, there was a shortcut out of the path of care.

On the shortcut to fame, argumentative cheating became a valid way of evaluating so that not complying with various rules of interpretive happiness became acceptable.

Among the ignorant, in the fertile land for ignorance, many other evils could be cultivated, misinterpretations were potentiated and Voltaire mocked these evils by highlighting the argumentative flaws of philosophers who wished to shed light on the great questions of the world with their digressions and analyses.

It is true that the Enlightenment, like all vain philosophy, is dissolved with truth and love manifested in a sincere life that does not support idealisms, because in the good work of cultivating beautiful gardens, only what is true prevails and vain digressions and analyses are naturally abhorred along with idealisms, as well as in the message of *Candide or Optimism*, let us work without thinking, let us cultivate our garden, a message adverse to the Enlightenment philosophy cradle of communism and socialism.

Candide or Optimism

The novel *Candide or Optimism* has thirty chapters where Voltaire uses the purity of his protagonist character, the young philosophy student Candide, against the supposed theory, present in the novel, of the best of all possible worlds by the connoisseur of various sciences Gottfried Wilhelm Leibniz, conciliator of the goodness of God, his infinite power and the necessary evil in the world in his work *Essays on Theodicy*.

The narrator of the proposed novel is omniscient in the third person, does not participate in the narrative and knows the whole story. He knows everything, including the characters' thoughts and gives us information in order to enrich rhetorical resources and their ideas.

Chapters 2, 3, 4 and 5 have enough elements to analyze the proposed irony phenomenon whose main target is Leibniz's optimism, however, parts of other chapters of the novel will be used to contribute to the critical analysis and coherent perception of the plot. The first chapter deals with the presentation of the initial environment, the characters and the main values involved. Up to chapter five, it is possible to exemplify the aspects of the resource of irony with its due critical analyses, without repeating them unnecessarily.

It is notable that the opposition between the character Candide and the phenomenon of optimism includes philosophy and Christian reason, which, in turn, in its history, encompasses Protestantism, so a large part of the ironies in the novel studied refer to the author's perception of philosophy and Christianity, as well as Christian morality and pertinent elements alive in the popular knowledge of his time, such as the tragic onslaught of the Anabaptist horde and the struggle against charity developed in Protestantism in the Modern Age.

By the ironic proposal, Voltaire threw ironies on these elements in his novel to contradict the impressions about them in favor of ridiculing them for satire and for evidencing hard and inconvenient truths.

Satire exists in a very complex and complete form and it exposes some inconvenient truths in some cases in parallel with the ridicule of what is excellent and venerable, Leibniz's conclusions about the best of all possible worlds.

The title of the novel '*Candide or Optimism*' proposes, at first, that there is an opposition between the representation of the character Candide and Leibniz's optimism.

The name 'Candide' denotes candor, that is, it has in its semantic charge the meanings of 'clear', 'objective', 'just', 'simple', 'naïve', 'pure' and 'innocent'; the conjunction 'or' indicates an alternation or exclusion between the first element of the title and what comes after it (the conjunction) which is 'Optimism' which refers to Gottfried Wilhelm Leibniz's theory of the best of all possible worlds, present in his work *Essays on Theodicy*, whose content shows that God could not have created but the best of all possible worlds, because if there was a way to create a better world, He, who is all-powerful, would have done it. Let us see Leibniz expound this reason in his great work:

> [8] It so happens that this supreme wisdom, united with a goodness that is no less infinite than itself, could not fail

> to choose the best. For as a lesser evil is a kind of good, so a lesser good is a kind of evil, if it hinders a greater good; and there would be something to correct in God's actions, if there were a way to do better. And as in mathematics, when there is neither the maximum nor the minimum, in the end no difference, everything is done equally; or when this is not possible, absolutely nothing is done; it may be said in the same way in matters of perfect wisdom, which is no less regulated than mathematics, that if there had been the best (optimum) among all possible worlds, God would have produced none." (Leibniz, 1710, p. 138)

The reference to Leibniz and his ideas found in the novel are worked by Voltaire with the use of several resources and the main one is irony in its form not useful for the evidence of the truth, which can serve satire, a combative form of irony, more than for any other phenomenon, however many other surrounding subjects are used in a true way to enrich the narrative.

Although sarcastic irony serves to evidence deception and, consequently, to evidence truth, in this case it serves to satirize Leibniz's ideas about the world being the best possible through semantic modification, although other elements have their real problems evidenced by the same resource.

Leibniz developed his essays based on the reason of the Catholic Church, exactly on the reasons expounded by St. Thomas Aquinas who was inspired by St. Augustine of Hippo about the permission given by God for evil to occur.

Although Leibniz is at the end of the complementation of reason, he was the victim of Voltaire's satire, possibly because he is an appeaser of the conflicts generated in Protestantism against Catholicism and attacking the appeaser or consolidator is of a genius that is found at a much higher level. It is like afflicting an individual who tries to break up a conflict so that the strife continues, and yet Voltaire had a clear side.

Leibniz (1646-1716) worked on the reason in the truth of the Catholic faith of St. Augustine of Hippo (354-430) which is, consequently, also the same as that of the Angelic Doctor St. Thomas Aquinas (1225-1274) regarding evil in the world where God would not allow an evil if it were not possible to derive a higher good from it, as we can see in the catechism:

> "The divine permission of physical evil and moral evil is a mystery, which God clarifies through his Son Jesus Christ, who died and rose again to overcome evil. Faith gives us the certainty that God would not permit evil, if he did not bring out of evil itself, in ways that only in eternal life will we know fully." (*Catechism*, p. 324)

This logical reason of St. Augustine and the Angelic Doctor St. Thomas, which is repeated in *Essays on Theodicy*, had its meaning modified by Voltaire to be capable of being satirized. In the novel, there is a modification in the semantic charge of the theory of the best of all possible worlds, where Voltaire replaces the definition of 'best possible' with another that does not allow afflictions around an individual with the intention of bringing playfulness to his work.

The Rhetorical Resource Irony

Voltaire does not touch Leibniz's theory, so he does not bring to his work the true theory, but another one that can be ridiculed.

The rhetorical device is used to elicit a reaction that can be laughter or loathing to something. In this case, the idea that the best of all worlds has no afflictions is abhorred.

To this end, Voltaire satirized the reasoning that leads to the deduction that this world is the best possible, by not bringing the complexity of theory to the novel, as well as it was evidenced that irony tended to despise the complex definition for the simple example, and the simple example was not enough to define.

God created this world and considered it good and regarding this statement, any element that contradicts the affirmation that there is no completeness (perfection) in it is a blasphemy that, before offending Leibniz, St. Thomas Aquinas or St. Augustine of Hippo, offends God by affirming that He has not finished what He has done, as if evil had no reason to exist in a finished world, however, this was not the intention of Voltaire, who sought the playful in his fictional narratives.

Basically, in the novel, the narrator's observations about the practical life of the protagonist Candide, who observes the world in the light of philosophy, contradict Leibniz's theory as if, in Candide's reality, in a world that contains evil, it could not be maintained.

By directing optimism in opposition to the character Candide, Voltaire intentionally uses sarcasm by indirectly calling Leibniz's theory confusing as something complex, difficult to explain and that is not maintained in the practical life of the characters, in fact that the practical journey of his character does not allow the existence of optimism as the novel conceives it and not as it really is in reality.

Candide, during the novel, questions the best of all possible worlds with ironic ancient Greek questions such as these: "Ah! Best of all worlds, where are you?" (*Candide or Optimism*, p. 16) and "While you were hanged, dissected, beaten and rowed in the galleys, did you always think that everything was going as well as possible?" (*Candide or Optimism*, p. 128), to develop a relationship of opposition between him, who is a character worked with characteristics in favor of truth, and Leibniz's optimism, to live up to the title of the novel that has in itself an **instrumental irony** if the extension of the definition of the phenomenon is considered.

This posture of Voltaire is a reflection of his condition of faith. Voltaire publicly took a more traditional and combative Catholic stance against Protestantism and Judaism.

The title informs that, in the development of the novel, there will be **situational ironies** for proposing that what is said about the world in the light of "Leibniz's theory" is not found where there is candor, that is, in the truth expressed in the practical life of Candide. In this way, **situational ironies** would be consequences of the irony phenomenon existing in some way as it was exposed.

Chapter two narrates what happened to Candide among the Bulgarians, how he engaged in the army, how he tried to disinherit, how he was flogged and how he obtained mercy from the king of the Bulgarians for being a philosopher.

Candide was called a hero while being flogged and forced to learn masterfully how to obey the formalities of the Bulgarian regiment. Let's see how this presents itself:

> "The next day, he did the exercises a little better and received only twenty blows. Two days later, he was beaten only ten times and he was looked upon by his companions as a prodigy.
> Stupefied, Candide still did not quite discern why he was a hero." (*Candide or Optimism*, p.11)

In the face of the inversions, Candide is described dumbfounded by the narrator, because the reason of the fictional world conflicts with his, which is clear and honest.

There is a clear **literary irony** related to the act of giving the title of hero to those who were practically forced to engage in war, that is, a satire against the heroism of those forcibly engaged in the army, as there is no heroism or individual desire for the act of engaging in the character they were forced to engage.

The definition of what it is to be a hero that we possess has undergone a single change. In the Helladic period, the hero was obliged to engage in the army and earned his glory for his deeds in war by plundering and pillaging. The greatest prize of this hero is to achieve immortality in the minds of the people who suffered his influences during his life as a warrior. In antiquity, Greek culture evolved a lot and the hero came to be defined by the examples of those who work to make the lives of others illustrious by pure altruism, such as the hero Theseus.

It is true that war, if used in favor of a good cause, works to make life illustrious, but we conceive heroism as altruism that is voluntary in favor of innocents, just as Candide also conceives, to characterize a semantic modification or inversion of the definition of what it is to be a hero, because what is said about being a hero does not represent what he is in the present of the narrative for Candide and also for who reads, which characterizes **literary irony** of the **verbal** or **instrumental types** and **dramatic irony** because we know more than those who call Cândido a hero. At this point, empathy between readers and Cândido is worked on, as a part of us may know about the diachrony about the hero phenomenon.

Let us note that irony at this level is difficult to detect because it is true that many do not know about the phenomenon of heroism in man to be able to understand it completely.

After being called a hero, Candide decided to desert and made use of philosophical reasoning to justify his act, as the narrator says: "Believing that this is a privilege of the human species, as well as of the animal species, to use his legs as he saw fit." (*Candide or Optimism*, p. 12), but he was overtaken by four other "heroes" and condemned to choose between being buffeted by the entire regiment thirty-six times or receiving twelve lead bullets in the brains at once.

In this case, there is more semantic inversion added to the definition of hero to laugh at the new definition. In this way, those who work to make the lives of others illustrious are not heroes, but the servants of tyranny.

Having questioning and thinking as a basis is what keeps philosophy that has always been well regarded in the search for truth and virtue in ancient philosophers such as Socrates, Plato and Aristotle. It so happens that, while humanity takes philosophical knowledge as a great source of theoretical knowledge that contributes to the practical, Voltaire, ironically, categorizes his philosopher characters as alienated from the reality of life without sparing even Candide, his protagonist, who, after having asked to be racked his brain, when he passed two out of thirty-six times through the whips of the entire Bulgarian regiment, He was acquitted by the king of the Bulgarians, who judged him to be alienated from the real world because he was dedicated to the study of metaphysics. Let's see how this presents itself:

> "... understood, from the information about Candide, that he was a young metaphysician, entirely ignorant of the things of the world..." (*Candide or Optimism*, p. 12)

The quality of being a student of metaphysics is parallel, equivalent and predicated with ignorance of practical knowledge in the thought of the king of the Bulgarians, which indicates that philosophical study is frowned upon by him, that he is the man given to practice, that is, he is to do, he has experience and Candide, for him, is to speak and not to practice.

Obviously, there is a **verbal irony** against those who study metaphysics that is a subdivision of philosophy and specifically Leibniz, author of *The Discourse on Metaphysics* and the theory of the best of all worlds. This **verbal irony** contributes to the satire against philosophers very present in this work.

As seen earlier, Voltaire did not fight metaphysics, but the unwary philosophers unwise to define metaphysics with their free reinterpretations, as Candide did as a naïve.

In this case, the irony lies in saying that those who study metaphysics do not understand the world, when metaphysics proposes to seek to explain the world with its experiences in such a way that this study transcends by encompassing all possible phenomena.

There is, obviously, a semantic change in the definitions of philosophical attributes in the speech of the king of the Bulgarians that contradicts the reality of metaphysics or its current definition and this semantic modification can be categorized as **verbal irony** or **instrumental irony**, as there is a distinction between the definition that the reader has of what is expected of philosophers who are students of metaphysics and the definition presented by the character king of the Bulgarians.

There is also an irony that is presented only from the reader's perspective, because what is said about metaphysics is not shared outside the text or in Candide's ideas as naïve.

What is said about those who study metaphysics is, in this case, inconsistent with what the reader possibly knows, which generates a **dramatic irony**.

The criticisms are directed at Leibniz and, obviously, if the target of these criticisms, specifically, were the Enlightenment philosophy, there would be no **dramatic irony** , but satire. It turns out that, indirectly, it is this way, even when developing an obvious reference to Leibniz, because the example of a philosopher is that of the philosopher of his time, the cowardly philosopher of the Enlightenment, which exemplifies the distortion of truth in the speech of the Enlightenment.

In chapter three, it is narrated how Candide saved himself from the Bulgarians and what happened to him next. The chapter begins with a battle between the king of the Bulgarians and the king of the abarians.

In the midst of the carnage, the narrator refers to the world as the best of them. Let's see how this presents itself:

> "Then the artillery removed from the best of all worlds about nine to ten thousand rascals that infested its surface." (*Candide or Optimism*, p. 13)

Obviously, the narrator wanted to create a contrast between what is narrated and the definition of the 'best of all worlds' in fiction, to create a **situational irony** where there is incoherence between what is or what is thought and what is said about the world as it is presented. With regard to the semantic inversion of the 'best possible', it is not possible to say that it generates **dramatic irony**. This is because the definition does not transcend from fiction to reality. Within the narrative, the definition of 'best possible' is true, although succinct and exemplified with the modified semantic load.

When the definition of the 'best of all worlds' of reality has been modified by a **verbal or** instrumental **irony** that does not include evil within fiction, there is **situational irony** in calling the best of all worlds a world that supports war while conceiving that the best of all worlds could not support evil.

Voltaire plants the semantic inversion that generates contradiction. It happens that many are those who accept semantic inversion as truth and, therefore, do not perceive the jocularity or irony.

Refined humor is difficult to understand, so a part of the literary texts designed to entertain often amaze and deceive readers unable to understand their playfulness.

Concepts that contradict each other are worked out by Voltaire throughout the length of his analyzed novel. There are ironies that consist of bringing an ironic consideration through the narrator. An example of this occurs when the narrator says that the soldiers of both armies were contaminating the best of all worlds with their presence, when Leibniz's theory clearly understands evil as part of it. Let's look at an excerpt that proves the acceptance of evil as part of the world in *Essays on Theodicy*:

> "For just as a lesser evil is a kind of good, so a lesser good is a kind of evil if it is a hindrance to a higher good; and there would be something to correct in God's actions if it were possible to do better." (Leibniz, 1710, p. 138)

We perceive Leibniz on the verge of explaining the unknown and incomprehensible, which can be very harmful to his own credibility.

The narrator of *Candide or Optimism* works on the idea that the best of all worlds has no equivalent evil, consequently using his own interpretation of Leibniz's theories that contradict Leibniz.

Humor can be evidenced and appreciated, in this case, in the narrator's considerations and not in what is narrated.

The Rhetorical Resource Irony

Certainly, when reading Voltaire, one is dealing with a very high level of rhetoric.

I raised some hypotheses about the reason for ironizing in *Candide or Optimism* when taking into account what I heard in orality and nothing they said seemed to me to be sensible, because they took fictional elements as a source of information for the real world.

Contrary to what they say, I realized that the narrator ridicules Leibniz's theory by altering part of its semantics so that it becomes susceptible to being satirized by literary art, which is in agreement with other occurrences and with part of the fruits of ancient Greek irony that serves to develop the playful, but this does not prevent us from considering other hypotheses at lower levels. Voltaire was complete in using rhetoric so that his readers could find elements that contribute to the rhetorical resource of irony on several levels.

In the passage in which the crossfire between Bulgarians and Abaros is narrated, Cândido's cowardice is evidenced in creating a link between it and his dedication to philosophical studies, in resuming and reinforcing the ironic idea that philosophers are inexperienced in ordinary and real life. Such a definition fits perfectly to the Enlightenment philosophers who were immersed in a quagmire of ignorance, however, in this case, we have the appropriate criticisms directed at any philosopher, that is, in the observance of his past reality among the Enlightenment, Voltaire defined a common character for the philosophers presented in his novel and with that satirized all philosophers.

Certainly Voltaire perceived the alienation from reality of the Enlightenment and this inspired him.

It is remarkable that he was actually referring to the Enlightenment, for there are no traces of

Christian devotion in the character of the philosophers who perceived debauchery as a common phenomenon and even as a necessary phenomenon.

The narrative voice uses the information that the phenomena occur in the best possible way to exemplify Cândido's cowardice in hiding from the carnage, when he says that **he hid in the best way possible**. Let's see how this presents itself:

> **"Candide, who trembled like a philosopher, hid himself as best he could** during the heroic carnage.
> Finally, while the two kings made the *Te Deum* sing, each in his own field, he chose to seek other places where he could reason about the effects and the causes."
> (*Candide or Optimism*, p. 13)

The image that the world forms about philosophers is like that of Socrates in search that leads to truth through virtue, so Voltaire uses the rhetorical resource of irony to contradict this by imputing to them that they are inexperienced and cowardly, by linking the definition of 'philosopher' with examples of his contemporaneity, the examples of the Enlightenment philosophers, which would create, secondarily, the "balance" that the modern resource lends itself to making in the ideas about the philosophers of his time according to many modern theorists.

As already said, balance is not the reason for the phenomenon to exist, nor is it a primary consequence.

The need for playfulness would justify the reason for the existence of irony that provokes laughter. Sarcastic irony, when it does not work for the sake of truth, can indicate only irrational hatred, which is certainly not the case.

On the run, after crossing piles and piles of bodies, Candide sought shelter in Holland, a country that had a reputation for being prosperous and inhabited by **people who preached themselves as Christians** (Protestants), which made him think that he would be welcomed and that he could return to live as in the past, when he lived in the castle of Thunder-tem-tronckn.

By narrating this yearning of Candide, Voltaire begins to work on an opposition between what Candide longs for and what happens in the **Dutch Protestant reality**.

The still naïve character Candide did not seem to know, having spent his life isolated from the world, in the castle of the Baron[8] of Thunder-tem-tronckn getting to know the world through the philosophy of Master Pangloss, that Protestants tended to kill each other for differences of ideas and that they were extremely miserly to the point of denying a piece of bread to a hungry man to justify avarice with the use of passages from the Scriptures and to the point of murdering anyone who said that charity could save. This generates a **dramatic irony** because we can know more than the character, which makes us know, before him, that he will be frustrated in what he thinks about the Protestants who preach themselves as Christians.

Pangloss taught metaphysical-theologian-cosmonigology, which may be an allusion to the person

---

[8] The predicate 'baron' has the acquisition associated with the purchase of it and Voltaire places the young Candide, who has generations of nobility in his blood, at the mercy of someone with a mediocre title of baron to disdain such a position, to disdain the purchase of nobility and dignity for little financial value and narrate the story of Candide with sides of nobility at the mercy of someone who bought the lowest position among the nobles with money and service entails irony.

of Leibniz who was a philosopher, mathematician, scientist, diplomat, librarian, and theologian.

We have Candide and Pangloss in the novel, just as we have Voltaire and Leibniz in reality, which characterizes an ironic composition of reality in a combative offensive fiction, that is, a satire.

In the novel, Candide had already consumed the reserves he carried with him and had to beg among those he thought were rich and charitable because they preached themselves as Christians, however, in Holland, Protestants killed people who preached that charity leads to heaven.

Intolerant of Protestantism and Judaism, Voltaire had a more traditional outlook for his upbringing and, influenced by St. Ignatius of Antioch, wrote to the religious study group SUR LES QUAKERS: "We **believe that those who profess an all-holy and spiritual religion should abstain, as far as possible, from Jewish ceremonies.**". SUR LES QUAKERS received Voltaire's attention after they realized that he was a constant victim of bad Christians and Voltaire was obviously one to influence and not to be influenced.

St. Ignatius (late first and early second centuries) fought the Judaizers and Voltaire did the same, developing a combative stance against the act of Judaizing that Pharisaism entails.

Voltaire highlights the malignity in the character who is a Protestant orator who spoke about charity, obviously in order to depreciate it as a vehicle that leads to salvation, and who did everything not to be charitable to Candide, even ignored his hunger and contradicted Christianity that has charity for the needy as its ideal.

With this, Voltaire created a **situational irony**, that is, a very believable situation because it is possible in the real world that contradicts the speech or thought of a character. Let's see how this presents itself:

> "He then addressed a man who had just spoken, alone, for an hour in a row on charity, in a large assembly. This speaker, looking at him sideways, said to him:
> - What are you doing here? Do you defend the good cause?
> "There is no effect without a cause," replied Candide modestly. Everything is necessarily chained together and ordered to the best of its ability. It was necessary for me to be removed from Kunigunda, to have to go through the whips, and it is necessary for me to ask for my bread until I can earn it. None of this could be otherwise.
> "My friend," said the speaker, "do you believe that the pope is the Antichrist?"
> "I hadn't heard of it yet," replied Candide. But, whether he is or not, I need a piece of bread.
> "You don't deserve to eat it," said the other. Go, rascal! Go, wretch! Do not bring me near to your life!
> The orator's wife, having put her head in the window and seeing a man who doubted that the pope was the Antichrist, poured on his head the whole contents of a..." (*Candide or Optimism*, p. 14)

The real grotesque intolerance of the Protestants, which made them practically expelled from France when they lost all civil rights, is very well portrayed in this passage, because Calvinist Protestants really hated anyone who thought differently from them who never had a unity of faith, which makes them divide into thousands of religious groups.

For example, Calvinism in Geneva burned Servetus of free religious interpretation in 1553, the first European to correctly describe the pulmonary circulation of blood, a remarkable medical contribution. Theologically, Servetus denied the Trinity and infant baptism, heretical positions for both Catholics and Protestants, which indicates an intolerance of Catholic rejection and estrangement and a Protestant murder.

Luther, for example, despised the letter of St. James and possibly disdained it when he called it a straw letter, because it spoke about love in the form of benevolence, charity, however, the Protestants portrayed were possibly Calvinists (more Judaizers) who reached avarice and atrocities far superior.

The speaker who spoke about charity ignored the hunger of his neighbor (Candide) and was concerned about protesting the Catholic Church by demanding that Candide confirm that the pope is the antichrist. The latter, the supposed Calvinist, entered into an argumentative dispute with Candide not to be charitable by giving him a piece of bread.

The eristics of the Protestant orator labored to conclude the reason that if charity does not lead to heaven, it may be abhorred.

In this excerpt there is literary irony of the **situational type** present in the relationship between the idea that Candide had about the rich Protestants called Christians by him and the reality that he lived with the speaker in the assembly who is one of the representatives of the thousands of types of Protestant faith, because what Candido thought was not confirmed in reality, it being exposed that the speaker endeavored to be miserly to him.

This ironic situation has its apex and reaffirmation when Cândido is treated in an extremely inappropriate way to the point of pouring the contents of a chamber pot into his head.

In addition to not being fed or welcomed under some Calvinist pretexts, Candide was treated in an extremely inadequate way. Exactly the opposite of what he considered happened, of what he expected from the Protestants he called rich Christians.

In writing *Candide or Optimism*, Voltaire made clear the distinction between Protestants and Catholics whose errors are evidenced.

The orator, a dissimulated character who spoke for an hour about charity, did everything not to practice it after finishing his speech about it and managed, through the divergence of ideas, to create a disaffection that justified his avarice, that is, Voltaire meant that he trained in solitude a lying speech full of deviations to lie about charity.

The argumentative game used by the speaker represents a lot of the routines and deviations that the bad Sophists used to try to win verbal disputes.

Let us remember that Socrates showed us that Euthydemus and Dionysodorus contradicted each other in order to be right and even said what follows when referring to them:

> "You sew men's mouths, as you also say; but because not only those of others, but it seemed also your own." (Plato, *Euthydemus*, p. 127)

This is the vice present in many arguers, which consists of using rhetorical inadequacies to be right in any way.

In this case, the orator entered into a dispute to gain the right to be covetous with Candide by being against Christianity without being called anti-Christian, and, if he defended charity during his speech, his right to contradict himself, but this was not the case, for it is known that Calvinists possess avarice in tradition.

The wretch in asking for a piece of bread to satisfy his hunger and the Protestant in denying him and demanding that he confirm the idea that the pope is the antichrist satirizes the motives that moved Protestants who forgot to serve God in love of neighbor and brings the nature of empty offenses against Catholicism in Voltaire's time, the partisan passion.

What Voltaire wished to express is that if Protestant heretical individuals do not wish to be charitable, in order to retain the titles of Christian persons, they must say that the pope is the antichrist.

For Protestants, it is convenient that they do not contradict the venerable ones for them, if possible even preach the antichrist that they retract, because they are Protestants and John Calvin who died by invoking the devil and blaspheming against God cannot be wrong in their perspective, which implies accepting or reinterpreting any absurdity of their life.

After being treated as a despicable and abominable creature for being in a situation of misery and having different and new ideas in relation to the ideas of the Protestants, Candide was unexpectedly welcomed by an Anabaptist named Tiago[9] who does not live in the countryside but in the city, has a trade and is charitable to foreigners. This description in itself is contradictory to the title of Anabaptist. Let's see his actions described by the narrator:

> "He took him to his house, cleaned him, gave him bread and beer, presented him with two florins and even wanted to teach him how to work in his manufacture of Persian fabrics that are manufactured in Holland." (*Candide or Optimism*, p. 15)

Throughout his novel, Voltaire narrates characters that have some title and are not what they represent in full or even in part. He does this by using the rhetorical resources available to create several **situational ironies** where there is no association between title and practical life in a partial or complete way.

In this case, he creates an irony with the standard of living of the Anabaptist James and that of the rich in the country of Holland, because the Anabaptists were originally peasants with the aggravating factor of having the ideal of an egalitarian society for those who did not belong to the horde whose application could be by imposition and consented among themselves to steal, the murder and destruction of other people's culture as justice, however the narrator, after calling the character Anabaptist, defines his activities as similar to those of free merchants, creating

---

[9] 'Tiago' is a form of the name 'Iaco' or 'Jacob' in German that is derived from the Hebrew 'Jacobus' which initially originated 'Tago' and which is used in Portugal and Brazil. It is equivalent to the name in the original French work 'Jacques', 'James' in English and 'Giacomo' in Italian.

a modern irony, because, if he was an Anabaptist, Tiago should be in the countryside, in the community helping in the fields or looting other people's goods and disturbing public order.

In this case, both the semantic change in the definition of being Anabaptist and the contradiction of the situation that James should be living in occur in some way, even without direct presentation in the text, which allows us to say that irony occurred or that the text is ironic.

The Anabaptists used, in an immanent way[10], faith, so they idealized revolutionizing society with their idealisms very close to the Enlightenment values that would form scientific socialism. They recruited the militants to die on the front line and Voltaire created an irony by developing the Anabaptist character James as a free merchant, someone who welcomes without observing whom to have some possible benefit by making the welcomed work for him in his trade that sells dreams and prestige combined with products.

Selling the illusion that carpets were provided from Persia may be a feature of free trade.

By enjoying commercial freedom, people sell and buy what they want, including sold dreams, which reinforces this **situational irony**.

Generally, when something is bought, the main reason is what the merchandise provides, that is, it can be something far beyond the product such as the illusion of having some product from Persia.

---

[10] 'Immanent' comes from 'immanence'. In Latin 'immanēns'; from 'im'-, 'in', and 'manēre', 'to inhabit', 'to remain'. It is a reference to keeping in or adapting to oneself. In this case, they adapted the faith to the Anabaptist ideology and kept it captive, molded.

Obviously, Voltaire wanted to impute the character of a free merchant to the Anabaptist when describing his product, by showing that he used a technique of deception in order to obtain more profit, which allows us to say that his situation is ironic if we take into account the breadth of the definition and the relationship between the Anabaptists' ideals of society and the character of the Anabaptist James.

To live up to his title, Tiago should have been taking care of the field in community, but he was taking care of the trade of mass-produced items itself.

Anabaptists, in the past, acted collectively out of idealism, so to impute to an Anabaptist individualistic character is ironic.

The name James may be an allusion to the disciple Saint James who says that religion immaculate before God is a set of charitable acts that culminate in practicing divine benevolence, which is somewhat ironic, because the Anabaptists were considered as a bloodthirsty horde of peasants who tried to impose their ideal of society and led their own to steal and murder, to completely contradict the charitable idea of St. James or the charitable character of the Anabaptist James in opposition to the Protestant preacher.

Armed with the Scriptures of the holy Catholic Church, the Anabaptists distorted it and protested against the doctrine of the apostles.

According to Fr. Julius Maria, author of *The Devil, Luther and Protestantism*, the Anabaptists tried to impose their ideal of society on the royalty supporting the Lutherans[11] who reacted violently by massacring them. (cf. MARIA, 1950, p. 64-99)

---

[11] 'Lutherans' is not a direct reference to a group called Lutherans, it is a reference

They were far from being charitable and desirous for the good of their neighbor like the charitable Anabaptist James. The Anabaptist Protestants formed a horde of peasants who broke into churches, broke images, destroyed altars, documents, and stole everything of value. The Anabaptists did this with more than a thousand convents and burned down more than three hundred Catholic churches. To this end, they deceived fanatics willing to die in the conflict for them, as can be seen below:

> "... they have penetrated the churches, pulverized the images and statues, dismantled the altars, committing all kinds of sacrilegious thefts; they razed more than a thousand convents; They burned down more than 300 churches and destroyed countless treasures of library manuscripts.
> It was the so-called Peasants' War, which spread to several provinces of Germany. In this fratricidal struggle more than 50,000 men died, deceived by the cruel Anabaptists, who sought to re-establish the republic, without civil power and without ecclesiastical authority." (Maria, 1950, p. 64)

They did not recruit to care for and give employment as the Anabaptist James did, but to die on the front lines of the conflict and at first at least thirty thousand Anabaptist men and servants of royalty paired with the Lutherans were killed and such a remarkable fact should be satirized by Voltaire by using irony.

---

to those who made choices in the faith after Luther was excommunicated for not repenting of having offended and lied against fellow Catholics.

The character of the welcoming Anabaptist who helps with the intention of doing good conflicts with the main action of the Anabaptists in history, which reinforces the ironic idea about the definition of the Anabaptist James.

The Anabaptists were slaughtered by a royalty paired with the Lutherans, that is, they were killed by other Protestants at the behest of Luther who wrote an open letter against the peasants to foment their massacre.

The existence of an Anabaptist living and living in a place taken over by Protestants also contradicts the situation established after the revolt of the Protestants who left the Holy Church, creating another irony, because it forms the statement that does not exist in the text, "The Anabaptist cannot live among the Lutherans and the Calvinists." when the Anabaptist had a business in Holland.

If the denial of the situation existed in the text, it would be **situational irony**, but I say that it exists in some unspoken way, which characterizes the ironic.

The Anabaptists acted against the institution that conceived faith and society differently from them, but the Anabaptist James welcomed Candide who differed from the Protestant orator with different philosophical and theological ideas and in this there is also a great irony, since it is historical that the horde of Anabaptists conceived the right to kill those who conceived society and religion differently from it.

Did the Anabaptist act charitably out of affront to the orator as a result of the Protestant revolt? Possibly, but he certainly did not act against his own impulses when he welcomed the naïve Cândido.

For a complete answer, just consult Voltaire about his creative process.

In chapter four, the novel narrates the reunion of Candide with his master Pangloss who was found in a deplorable state.

Pangloss tells Candide of numerous misfortunes caused by the Bulgarian soldiers, tells how he contracted syphilis and reached his pitiful state. When narrating how he contracted syphilis, he defines the genealogy of the disease by citing in it some representatives of the Catholic Church who should be celibate and heterosexual as adulterers, homosexuals and vehicles that transmit the disease. Let's see how this presents itself:

> "Paquette had received this gift from a very wise Franciscan friar, who had gone to the fountain. For he had obtained it from an old countess, who had received it from a cavalry captain who owed it to a marchioness, who had taken it from a page, who had received it from a Jesuit[..]" (*Candide or Optimism*, p. 17, 18)

Again, Voltaire disentangles, through the rhetorical resource under study, titles and responsibilities, since a very wise Franciscan friar who should be celibate, penitent and of monastic life is narrated as a man who seeks sexual pleasure and who fornicated with Paquette to transmit to him the venereal disease acquired by having fornicated with a countess.

Nobles, subordinates of nobles and members of the Church in evidenced worldly acts. Voltaire set out to create a notion opposite to the posture of a priest and a situation that does not suit the character of a priest. In praise of the Franciscan friar's wisdom, Voltaire was certainly referring to his cleverness and success in seeking to enjoy the favors of women, thereby creating an **instrumental irony** that works on the definition formed about men with ecclesial authority in truth that wisdom is knowing how to make use and supposed good use of phenomena.

This irony occurs only between the narrator and the reader, as it is notable that the characters are not surprised or wish that priests are celibate and heterosexual.

There is also not a single hint of reverence for ecclesial authorities, which reinforces the idea that the examples of philosophers used were those close to those of the author Voltaire, that is, the Enlightenment.

This is part of the creation of a new reality through fiction that takes certain libertine values as a rule, and this idealism is part of Enlightenment philosophy.

Next, Pangloss, very unfortunate by syphilis, tries to refute the idea that syphilis is an evil, arguing that, if it weren't for it, we wouldn't have chocolate or cochineal.

Voltaire, at this moment, ironizes the following statement of St. Thomas Aquinas: "God would not allow evil if he did not bring out evil itself." (*Catechism*, chapter IV, paragraph 324) repeated in a distinct way by Leibniz to give the reason for the coexistence of good and evil in the world in his essays.

It so happens that Voltaire believed that St. Augustine's statement and tradition were enough because it was not possible to understand it at the level of developing essays as Leibniz did.

The irony is in saying that having chocolate and cochineal, which provide more pleasure than good nutrition, is more important or is a greater good than the evils brought by syphilis, when no human being in his right mind would judge in this way.

This fragment places the reader in a condition of wisdom superior to that of the great philosopher Pangloss, which precedes the occurrence of **dramatic irony**.

The irony occurs in ideas, in the common criterion for judging phenomena, in Pangloss's conclusion based on Leibniz who was based on St. Thomas who was based on St. Augustine.

Pangloss, besides being treated for his illness, was admitted as an accountant by the Anabaptist James and, on a business trip, while he was lecturing on the production of good by private sufferings, the sky darkened, the wind began to blow stronger and a storm began, which would be a misrepresentation of Leibniz's theory by Pangloss if his conclusions were not fictional, by using one's own measure to measure[12] the world. Because it is fiction, what Voltaire says is that the theory of the best of all possible worlds is incomprehensible, which makes it liable to misrepresentation.

After the storm started, there was an earthquake that toppled the buildings and crushed more than 30,000 people, which contradicts the theory of individual suffering that Pangloss had just defended, as it was a generalized suffering which generates a **situational irony**. Let us see Pangloss setting out to

---

[12] Using one's own measure or one's own example to measure the world is what characterizes idiocy with the Greek semantic load and such a state was acquired with social isolation that made idiotized people create their own definitions about the world.

explain individual sufferings as promoters of a general good before the tremor:

> "All this was indispensable," replied the one-eyed doctor, "and particular sufferings produce the general good, so that the more there are particular sufferings, the more all is well." (*Candide or Optimism*, p. 19)

Even if thirty thousand deaths were considered as individual sufferings, what happens after the earthquake does not resemble a good at all, which contradicts the speech of the *great* philosopher Pangloss and evidences the existence of **literary irony** of the **situational** type in some way.

In this case, Voltaire works with the jocular idea that the philosopher talks about what he does not know and that he speaks to justify the ideas he has previously assimilated.

Ironically, thirty thousand was also the approximate number of Anabaptists killed by Martin Luther's command, which, if added to those who had died on the front lines of combat, exceeds sixty thousand.

Pangloss could not say that his individual suffering had brought about any good in the observance of reality and yet he did.

To speak in an attempt to perceive the balance between good and evil can culminate in error, which is in complete coherence with the conclusions of Cândido matured, at the end of the novel, when he says, together with his followers, that they should take care of their own garden and forget about ramblings and analyses.

With the agitation of the sea and several disputes on the vessel, the Anabaptist James falls from the vessel into the sea, is immediately engulfed and Pangloss uses the idea that everything went in the best possible way to justify his own cowardice for having prevented Candide from throwing himself into the sea to save the Anabaptist James who had saved and employed him. Let's see what the narrator says about this:

> "He (Candide) wanted to throw himself after him (the Anabaptist James) into the sea, but the philosopher Pangloss prevented him, proving to him that the inlet of Lisbon had been formed expressly so that that the Anabaptist would drown in it." (*Candide or Optimism*, p. 20)

Philosophers are attributed to acting contrary to heroic action characterized by the ability to give one's own life to make the lives of others illustrious, that is, in the narrator's conclusions, philosophers are within the semantic field of villainy and have sensible conclusions to act against heroisms out of cowardice.

Voltaire rightly observed, because history is full of people who talk a lot and who, when they are called to work for their supposedly high values displayed only by giving an opinion, run away from the job. There are many experts in education full of idealism and opinions about the lives of others who are unable to educate their own children. True eunuchs of arem who see how their master does it and know about exactly everything, but do not have the tool to do it and who, because they know how to write and read, even without good proficiency, try to compensate for the lack of virility by trying to advise the world just like Pangloss.

When the narrator says that Pangloss proved to Candide, he was obviously ridiculing the way in which the theoretical and philosophical knowledge of bad philosophers is given. He meant that Pangloss's proof was unproven in the reality of the characters. This irony calls into question the effectiveness of philosophy in the search for truth in Leibniz's theory, because the narrator's speech entails the truth that when philosophers, of the exposed type (the misrepresenters), say that they prove, what they offer is not a proof, but a deception or an argumentative trick in order to work for their own lazy and cowardly ends. There is also a **situational irony**, on the part of the narrator, when he writes 'proving to him', because the proof does not happen in the narrated situation. Obviously, a proof was not what Pangloss offered to Candide, but a fallacy, a sophistry, a routine of cheating, to deceive.

What Pangloss said was in accordance with the already exposed tendency of philosophers to cowardice and did not generate the good fruit of altruism, benevolence for the benevolent convict. He could say that the Anabaptist was unsaved, but he rambled philosophically about what happened to make philosophy null and void in fiction, another unnecessary and mistaken way of describing an observable phenomenon, or even of lying unnecessarily.

At the end of the novel, Candide, already completely incredulous in relation to the theory of the best of all possible worlds in the way his master presented it, raises the following question to Pangloss: "When you were hanged, dissected, covered with blows, and rowed in the galleys, did you still continue to think that everything was going in the best possible way?" And Pangloss replies to him as follows:

> "... for I am a philosopher: it is not convenient for me to say myself. Leibniz can't be wrong." (*Candide or Optimism*, p. 129)

In Candide's questioning, we have a sample of ancient Socratic Greek irony that leads the interlocutor Pangloss to take a different conclusion or contrary to his own convictions and, later, Pangloss wishes to keep the title of philosopher at the price of believing in a "lie", by contradicting the reason of philosophy, which is to explain the truth, as if philosophy had the function of justifying false ideas, which characterizes ridicule, derision or mockery to compose satire against the distorting philosophers, the trickster philosophers of their time.

Voltaire portrays the character of foolish philosophers, those who talk about what is ineffable, indefinable, as alienated from the truth of life when he places Pangloss in a situation where he must prove an idea of his that is not maintained in his practical life by the title of philosopher.

In this case, for Pangloss, what characterizes him as a philosopher is to have a set of ideas by assimilation without necessarily reasoning about it, that is, he must think in a certain way to be predicated as a philosopher in a state of blind faith.

It is precisely when the character Pangloss makes reference to Leibniz that the great ancient Socratic Greek irony worked throughout the novel is confirmed with all the necessary elements to converge in a single point, a perfect irony in the mold of Socrates' ironies.

It is logically concluded, since it is true that the theory of the best of all worlds is not liable to be criticized through simple questions such as those presented in the novel, that Voltaire created a new definition or *straw man*[13] of Leibniz's theory, by modifying the meaning of 'optimum' (best possible), which is observable in Pangloss's conception before his suffering on the best of all worlds to create an **instrumental irony** that helps in the creation of the best of all worlds. of **situational ironies** throughout the novel with the intention of contradicting or getting laughs out of a subject that is serious and complex.

When working for the Anabaptist James, Pangloss already accepted sufferings in a superior way as phenomena present in the world, but he could not contradict himself, because he would not be respected as a philosopher and thus all philosophers of this type are caught in errors and lies, because, when trying to justify themselves, they get dirty, make mistakes and contradict themselves in a superior way during the process.

If we understand the novel as an integral body, it is clear what has been said above about the additions to the phenomenon studied, it is clear that the scholars of the phenomenon have observed the necessary causes for the irony phenomenon to exist and its consequences to define it, which allows us to say that a text is ironic when there is implicit irony that occurs in some way.

---

[13] In the case exposed, the word '*straw man*' comes from **the straw man fallacy,** which consists of an argument in which the position of the interlocutor (Candide) or figure not present (Leibniz) in the debate is ignored, which is replaced by a distorted version (a straw man) that it misrepresents.

Even Candide, convinced that there is necessary evil in the world, does not show that he understands that, in a complete world, there is also evil, which can lead to understanding that, in his conception of the perfect world, evil does not exist.

Leibniz accepts the afflictions of the world as part of it, however, in the novel, the theory is presented as if he said that, because the world is complete (perfect), evil should not exist and if it does exist it serves to lead to a higher good. This is because our conception of 'perfection', which should be the same as 'complete' because we live in this reality, is the one we have of the completeness of the Christian paradise when the completeness of this reality is different from that of paradise. Even when the characters are forced to recognize evil as a service for a higher good, they are totally or partially contradicted by the narrated facts developed by the author who uses rhetorical resources to work out these contradictions.

In the novel analyzed, the rhetorical resource irony allowed the creation of literary art by altering meanings and creating distinctions between what is and what is said about something, exactly as the ancient philosophers and sophists did by provoking laughter by ridiculing and modifying through situational inconsistencies and fallacies that appear rational and that the resource allows to create in fiction.

For so many ironies to be maintained, it was necessary not to have many conditions for interpretative success from the perspective of the characters, when they dealt with the phenomena involved and their complexities, which does not occur in the same way from the perspective of the reader who can be educated, appreciate the literary beauty and entertain himself.

At the end of the novel there is a very clear message of rejection against digressions, analyses and relations with foreign powers and in favor of valuing work, of taking care of one's own life, of taking care of one's own garden. A message contrary to what was developing among the philosophers of his time, socialism and communism.

The character Candide, while naïve, accepted everything she heard from Pangloss's philosophy and longed for Kunigunda's love. In the end, when wise, he questioned his master and made use of ancient Greek irony willing to marry Kunigunda out of obligation to do him a favor. The idea, in this case, is that passions are the property of naïve people in a superior way and they prevent us from becoming wise. There is also the notion that we may not know what we really want and that digressions and analyses are incapable of providing a complete definition of the phenomena of the world.

In the end, Candide's knowledge becomes enviable, which makes us, readers, desire or assimilate part of his values and certainties, which is to oppose Leibniz's supposed unreal optimism.

Jonas Batista dos Santos

Memnon or Human Wisdom

Due to the physical limit characteristic of the short story genre, the narrative voice should inform directly to only a necessary part of the story and, when possible, use resources to bring a new deep reality in time and in a proper place *that is not current life and not real life* to respect the reality that certain mysteries are more credible when they are not explained, Thus, by not having the origin or the detail of the origin of some phenomenon, it is possible to generate a depth effect in the fictional text and this is important in a literary text that has a physical limit of around thirty pages such as the short story, however, the lack of information should not be vitiated, the lack of information should have an artistic purpose and not to deceive readers or interlocutors, for, as has already been stated in a definite way, the motive allows phenomena to manifest themselves in different ways.

In the narrative text short story there is not enough space as in the novel that allows the author to work the plot in the midst of merely illustrative items that only contribute to aesthetics, so in it, the author delivers some information in a succinct and objective way as Voltaire did in *Memnon or Human Wisdom*, when he wrote that *Memnon one day conceived the foolish project of being perfectly wise.*

Due to the physical limit, a good short story has a lot of depth of meanings, so that the dissertation on it can exceed its volume several times, if it is analyzed completely. The hidden origins of ideas can bring, in addition to a broader fictional reality, the desire to know more or bring uncertainties to some present elements that can be interpreted on many bases, which is why uncertainties are common when analyzing short stories.

In the first paragraph, the omniscient third-person narrative voice implies that his protagonist will commit a fault or a series of them against what is estimated as common sense in the text, as he calls his protagonist's project foolish and, later, madness.

The tale ironizes the struggle against passions and vices by making use of the will, so that the protagonist promises a series of sensible changes for himself, but is led to practice the vices that he had promised not to practice soon after the promise.

Every sinner who has used only his own strength to leave sin should feel empathy for Memnon, for he has the same problem as his not resisting the temptations of the world.

The proposal also highlights the irony of philosophers who, like Memnon, cannot manage their own existence and consider themselves able to advise on behalf of other people. This proposal may lead, at first, to the obvious conclusion that the tale deals with the protagonist's misfortune, but, due to the depth of the text, several speculations are appropriate.

Memnon promises himself that he will never love any woman, that, when he perceives a perfect feminine beauty, he will see how the woman will look at the end of old age and that he will also no longer turn his head when admiring women, as it is possible to see:

> "To be wise enough, and therefore happy enough," considered Memnon, "it is enough to have no passions; **And nothing is easier, as we know**. First of all, I will never love any woman: for,

when I see a perfect beauty, I will say to myself: "These faces will wrinkle one day; those beautiful eyes will be edged red; those stiff breasts will become sagging and pendulous; this beautiful head will lose its hair." Just look at it now with the eyes with which I will see it then, and that head will not turn mine."

These promises and others related to moderation in financial spending were made by Memnon inside his room, that is, without contact with the society of the character's fictional reality, without being in front of the threats that Memnon tried to avoid, in a moment of philosophical reflection on what is the best thing to do.

Then Memnon had just promised himself not to see perfect beauty, put his head out the window and saw the beauty of a young woman, which characterizes an irony that works against those who promise moderation for themselves.

The one who informs the reader that Memnon sees the beauty of a young woman is the omniscient narrator by evidencing the inability of a man to promise himself something contrary to his own drives, Memnon does not admit it, because he seeks to hide his person from himself.

The situation presented by the omniscient narrator who informs that Memnon saw the beauty of a young woman contradicts Memnon's speech that he promised not to see beauty, which characterizes **situational irony**.

He saw Memnon, accompanied by an old woman, a young woman whom he thought very beautiful. She was sighing and crying, and in order to deceive himself, Memnon used as a worthy justification the consolation of the young woman, which characterizes self-deception and could precede **dramatic irony**.

Because he saw what he saw and did what he did, he involuntarily undid the promise of not seeing female beauty in the moment after it, as it is possible to see:

> "Having thus made his little plan of wisdom inside the room, Memnon put his head to the window. He saw two women walking under the plane trees near his house. One was old and didn't seem to think about anything. The other was young, beautiful, and seemed very worried. He sighed, wept, and with that he only increased his graces. Our philosopher was impressed, not by the beauty of the lady (he was sure not to indulge in such weaknesses), but by the distress in which he saw her. He went down into the street and approached the young woman, intending to console her wisely."

The narrator informs that Memnon's intention is to wisely console the young woman as if to say the opposite of what he thinks, as if to ironize, because he had shown that the young woman's crying only increased her graces that are meant by her beauty, that is, Memnon was being carried away by the beauty of the young woman all the time.

Every man wants to be a hero and helping a young woman would be a greater attraction.

The Rhetorical Resource Irony

It is true that opportunistic women pretend to be victims so that fools will supply their needs by thinking that they are committing worthy acts of heroism.

Every man desires to tell his deeds of heroism. Let's look at these qualities in the hero Theseus:

> "I was not surprised if you prolonged your speech by rejoicing with your daughters, or if you preferred their words to mine. None of this matters to us, for we are occupied with making life illustrious more with deeds than with speeches. And I prove: from what I swore I have not deceived you in anything, old man, for here I am and I bring you safely, untouched by threats." (Sophocles, *Oedipus at Colonus*, 1140-1145 p. 79)

The fragment on display expresses a value that is very present in the works of Voltaire and Plato, which is to make life illustrious by effect and not by words. In Euthydemus, Socrates criticizes those who make philosophy the only path to excellence. For Socrates, the wise man is the one who makes correct use of phenomena, which implies the existence of philosophy as a support for provision and not as a provision as Memnon conceived.

As seen in the speech of the king of the Bulgarians, Voltaire brings the same value when he calls the students of metaphysics alienated to the world.

The philosopher, in this universe, because he has the characteristic of cowardice, seeks to help with wise words when there is no one who is consoled by only listening to our wisdom.

Let us note that the philosopher, in this tale, is an individual who considers himself able to advise and console with his wisdom, but cannot manage his own life when he is led to meet the two women by the beauty of the young woman at the price of abandoning without realizing what he promised to himself and this truth that contradicts has part of himself that is not physically expressed as well as an ironic questioning that leads to a truth that can contradict.

Voltaire, as in his novel *Candide or Optimism*, works on the image of the philosopher as someone unfit for advice and in this there is a great irony that touches on the conceptions formed about philosophers, a semantic inversion of the definitions involved that contributes to satirize philosophers seen as wise advisors.

Obviously, beauty exists regardless of whether one wishes to see it or not. In woman, beauty is conjoined with fertility and signs of physical health that the old woman no longer possessed in *Memnon or Human Wisdom*.

It is not possible to use at will against will as Memnon tried to do, that is, to use desire against desire or reality, in truth there is desire in vices and, by this exposed truth, it is possible to understand why idealisms against nature do not work completely.

Being a philosopher and having reflected many times on himself and the world, Memnon understands that abandoning addictions is the initial act for there to be an improvement in his quality of life, but he, like most people, cannot help making mistakes.

In this tale, the abysses of the proposal to be perfect are endless, one evil brings another evil and the beautiful opportunistic lady leads Memnon to be robbed.

Embarrassed and desperate, to console himself, Memnon decides to go out to dinner with some close friends in order to forget the foolishness he committed, he promised himself that he would drink moderately, but he contradicted himself with the acts by committing excesses and a misfortune happened to him, as it is possible to see:

> "If I am alone at home," he said, "my mind will be preoccupied with my sad adventure, I will not be able to eat, and I will end up getting sick. It's better to go and have a frugal meal with my intimates. I will forget, in the sweetness of your conviviality, the foolishness I did this morning." Attends the meeting; they find him a little taciturn. They force him to drink to dispel sadness. A little wine taken in moderation is a remedy for the soul and body. This is how the wise Memnon thinks; and gets drunk. Then they propose a game to him. A little game between friends is an honest pastime. He plays; they earn everything they have in their purse, and four times as much on their word. In the middle of the game a dispute arises; Tempers flare: one of his close friends throws a glass of dice in his face and pours out an eye. They carry home the wise Memnon, drunk, penniless, and with one eye too few."

The narrative voice emphasizes calling Memnon wise, being wise who makes good use, who acts or speaks in accordance with reason and morals, acts with prudence and life experience, sensibility and balance and Memnon cannot be predicated with such virtues, which characterizes the affirmation of the narrative voice as ironic because there is semantic modification.

The narrator's lyrical self creates a semantic modification between what is known about being wise and the definition presented in the tale, he also calls Memnon wise while practicing imprudence and lack of moderations, which characterizes **situational irony** to be maintained by semantic modification.

The way to the solution of the problem of addiction is not found in this tale because it deals in a real way with the deception of most people or because the lyrical self takes as a solution moderation instead of the complete aversion to vices idealized at first, as is perceptible in the course of the narrative, and the philosopher would not be the solver of the problem because he is incapable of managing his own existence.

The real solution is far from a promise to yourself not to make mistakes. The way to the solution of vices is, in fact, outside the fictitious text, in denying oneself. Statistically, those who managed to leave the vices, sought the sweet lap of God through prayer or were struck by a disease that took away their pleasure from the vices and, in the tale studied, the man who is in search of what he thinks is the best for himself, conceives that pleasure is the solution and receives an injury to the eye, which makes him an injured person closer to the wretched who sought Jesus who told them to go and sin no more.

It is remarkable that almost all those who repent and who are narrated in the Scriptures were in great misfortune and Voltaire of Catholic education used this reason to bring disgrace to his protagonist, which leads us to think: "Is Memnon now going to moderate his actions?"

In this case, the logic that allows the deception to be renewed is that, if the human being seeks the best for himself in personal reasons and takes vices as something good for himself, then there is no way not to make mistakes in this way, in truth that, for him, the error is the best for him, even without conceiving it as the best for oneself when making use of reason.

In the end, the author, through a dialogue between Memnon and a celestial creature, creates a relationship between exaggeration by moderating customs, misadventures and madness as it is possible to understand:

> "Your luck will change," said the animal in the star. – It is true that you will always be one-eyed; "but, apart from that, you will still be quite happy, as long as you do not make the foolish design of being perfectly wise." [...]
> [...]one has less wisdom and pleasure in the second than in the first, less in the third than in the second. And so on until the last one, where everyone is completely crazy.
> "I am very afraid," said Memnon, "that this little earthly globe of ours is precisely the madhouse of the universe of which you do me the honour of speaking. [...]

> [...]"Ah! Memnon exclaimed. "It is clear that certain poets, certain philosophers, are not right in saying that all is well.
>
> "On the contrary, they are quite right," replied the philosopher from above, "taking into account the arrangement of the entire universe.
>
> "Ah! I will only believe that," replied poor Memnon when he is no longer one-eyed."

Let us note that, in the story, the solution to addictions is not obtained by denying one's own desires, but by avoiding wanting to be the best, as if a cosmic energy worked against those who want to achieve perfection. This force would aim to establish a "balance", obviously because it is a playful text.

Inspired by the wretches with God's consent of his tradition, Memnon suffered something like a divine test to prove his worth as a sage, that is, by promising to be extremely wise, Memnon would have attracted to himself a test.

In apparent disgrace, Memnon began to doubt that everything was going for the best possible and the ironic reference to Leibniz is clear enough to identify him in this surreal and unexpected ending where it is noticeable that the protagonist Memnon imposes the condition of only believing in the words of the celestial being if he has his pierced eye restored, despite recognizing a certain truth in balance and having received the revelation of a celestial being who came to explain to him the past, the present and the future.

The richness of the tale has a lot of depth, so that it allows us to speculate whether there was something so fantastic in the narrative text or whether

Memnon slept, had a fever from infection in his pierced eye and was delirious in his philosophical reason that could not bring him more than a comforting dream, a deception.

I believe that, due to the depth of the genre, both forms coexist.

It is noted that the great irony lies in considering that it is possible to modify one's own reality and be contradicted by oneself and by a cosmic force present in everything that modifies the situation, to prevent excesses and to generate a balance to the point of severely punishing those who wish to work or severely resist against their wills.

This ironic message shows the same lack of control over the causes and consequences seen in the novel *Candide or Optimism* and the smallness of the human race in the face of phenomena, which includes drives.

History of Escarmentado's Travels Written by Himself

The narrator created by Voltaire presents Escarmentado, the offspring of Candia who received verses from a mediocre poet whose content narrated his story as if he descended from Minos and who, after his father fell into disgrace, received other verses as if he descended only from Pasiphae and his lover.

Minos is a legendary demigod from Greek mythology and Pasiphae is one of Helios' daughters, very jealous who, to prevent her husband's union with other women, placed a curse on him. In the curse, the women who loved him would die devoured by snakes, scorpions, and centipedes that would come out of his penis.

It is evident that French culture is a consumer of Greek culture and knowing about the pertinent cultures allows us to open a certain range of interpretations regarding the misfortune that befell Escarmentado's father. Let's see how the tale is introduced:

> "I was born in 1800 in the city of Candia, of which my father was governor. It reminds me that a mediocre poet, who was not mediocre harsh, composed some bad verses in my praise, in which he made me descend from Minos in a straight line; but when my father was disgraced, he wrote other verses, in which I was descended only from Pasiphae and his lover."

Regardless of where the depth of this tale may lead, we must recognize its mastery. Through the song it is possible to speculate that Escarmentado's father discovered that he was betrayed through a venereal disease and that the son was not his. There are many possibilities beyond these.

The satirical irony lies in changing the immutable that is a genealogy, in abruptly withdrawing honor because it is linked to social prestige and in the mouth that praises that can easily be defamed.

Voltaire, in his current life, questioned social positions and did the same in this text.

Through the rhetorical resource, irony removed the meaning or reason that links title, social prestige and honor to a stability to speak of a jocular instability as a madness worthy of laughter present in the world.

It is remarkable that every author brings phenomena he knows to his fictions and it is no coincidence that the following excerpt partially portrays Voltaire's childhood:

> "When I turned fifteen, my father sent me to study in Rome. I came hoping to learn all the truths; for I had hitherto been taught the exact opposite, as is customary in this world, from China to the Alps. Monsignor Profondo, to whom he had been recommended, was a singular man and one of the most terrible sages that ever existed in the world. He wanted to teach me the categories of Aristotle, and he was on the point of placing me in the category of his mignons: I escaped in time. I saw

processions, exorcisms and some rapine. They said, but falsely, that the signora Olympias, a person of great prudence, sold a lot of things that should not be sold. **I was at an age when all this seemed very fun to me.** A young lady of very mild habits, named Signora Fatelo, was pleased to love me. I was courted by the Reverend Father Poignardini and the Reverend Father Acomiti, young teachers of an order that no longer exists: she put them in agreement, granting me her graces; but at the same time he was in danger of being excommunicated and poisoned. So I left, very happy with the architecture of S. Pedro."

After his father tried to remove him from the clutches of his recruiters, the Illuminists who made him a member of the Temple Libertines, Voltaire noticed some examples of debauchery among those who should give him a high education and this is the inspiration of the text.

What he said is far from empty speculation. Bringing the truth of one's own life to literary works is an act present in the works of many authors. I dare say that everyone practices this.

They not only deposit what they know about the world in their productions, they also deposit their own personal aspirations that influence the creativity of the narratives.

It is remarkable that a writer without virility can possess the greatest mastery in describing leather bags with cufflinks and tattooed men's bodies. He develops his protagonists who do not develop any affection for their supporting actors in a healthy way, even though they live long plots that force them to spend a certain amount of time with them, as when I once read a very well written detective novel whose protagonist is a eunuch expert in symbology, as if he were a much improved copy of the author.

Much has been said that one cannot link the personal life of an author with his work, but I think that, if this were valid, they would not put the truths of their own life together with their reasons in their fictions.

I think that truth and passion are the elements that make fictions good.

I have a few books on literary analysis that tell me not to look at what I see. On the other hand, I think that, at a certain point in an analysis, every element is valid, including speculative ones, but before developing analysis within the level of speculations, it is necessary to take the trouble to analyze in a formal way.

Mocked, our protagonist travels through France during the reign of Louis XIII called by the voice he narrates the just. In this land, he was offered, wherever he went, portions of Marshal d'Ancre.

This marshal, from the narrator's perspective, was unjustly killed by Louis XIII, creating a semantic inversion by defining fair and a **situational irony** for the injustice being narrated in two acts of the king after he was called just by the voice he narrates.

It is remarkable to play with different perspectives that contradict each other, which is something common in Voltaire's productions and characterizes the ironic text.

Our narrator demonstrates that he is aware of isolated acts practiced by representatives of the Catholic Church with the purpose of avoiding violence through it:

> "I passed to England: the same quarrels excited the same furors there, Catholic saints had resolved, for the good of the Church, to blow up gunpowder, the king, the royal family, and the whole Parliament, and to rid England of such heretics. They showed me the place where Blessed Queen Mary, daughter of Henry VIII, had more than five hundred of her subjects burned. A priest assured me that it was a beautiful action: first, because those whom they had burned were English; secondly, because they never used holy water and did not believe in St. Patrick's hole. He was astonished that they had not yet canonized Queen Mary; but he expected it soon, as soon as the cardinal-nephew had some leisure."

It is facetious to mention violence without informing which side it belongs to and to equate the violence of the good with the violence of the bad.

To say that using violence to avoid violence is ironic, however, a necessary resource, because it is necessary to act against the violence of the wicked.

In the midst of jocular justifications, it is also said that representatives of the Church killed heretics because they were killing. This is one of many confusions between the fictional and the real to bring humor with the use of **situational irony**, because the truth is that, in part of the Middle Ages, some criminals were judged by secular courts and it was certain that they would receive a death sentence, which made them cry out to be judged by the Holy Inquisition, to this end, they pretended to be demon-possessed and blasphemed against God to be judged by an ecclesiastical court in the hope of receiving an acquittal.

To highlight the sanctity of the holy Catholic Church is not facetious, it is facetious to put it in a situation of error or to highlight the traitors as its representatives, however, after the pact of verisimilitude of fiction is broken, traitors are traitors and representatives are representatives.

Voltaire understood human motives and knew that if we created an association called The Righteous, the unjust would be in it, among the righteous people to obtain some advantage such as pretending to be righteous people.

When heading to the country of Holland, which is a land considered Protestant, a sewer of Catholicism at that time, in order to find a calmer people, our narrator tells us that, in it, the people killed for dogmatic differences and one was being killed by Protestants for believing that charity saved the soul.

Catholics killing on one side and Protestants murdering on the other. The narrative speaks of genocide as a custom of the human race that exists regardless of their convictions. Let us observe the reason of the Protestant preacher who condemns the charitable to death:

> "He has done much worse," a black-robed preacher replied. – This man believes that one can be saved by good works, as well as by faith. He sees that, if such opinions were in force, a republic could not subsist, and that there is need of very severe laws to suppress these scandals."

The Rhetorical Resource Irony

As in *Candide or Optimism*, there is a **situational irony** that involves what is thought about Holland and what happens in it that is inspired by the reality that entailed killing for disagreements promoted by Protestants.

The sharing of ideas between novel and short story is noticeable, as a similar episode is narrated in the novel *Candide or Optimism*, where a Protestant speaker preached against charity.

Our protagonist leaves Holland and goes to Spain because he did not perceive any attraction where he was, but, after appreciating the beauty of the land of his destination, he witnessed the following:

> "Then came an army of priests, in formation of two, white, black, gray, shod, barefoot, with beards, without beards, hooded, without hoods; then the executioner marched; then, in the midst of the Alguazis and the grandees, there were about forty people dressed in sacks, on which they had painted devils and flames. They were Jews who had not wanted to renounce Moses, Christians who had married their wives, or who had not adored Our Lady of Atocha, or who did not want to give up their affairs in favor of the hieronimite brothers. They devoutly sang beautiful prayers. then they burned all the culprits with a slow fire, with which the royal family seemed extremely edified."

The narrative expresses the same acts and values attributed to the inquisitorial trial present in the novel *Candide or Optimism* when Pangloss and Candide are arrested and convicted.

Even the characterization of the characters is the same at the time of condemnation, because, in both narratives, devils and flames are painted on the clothes of the condemned.

Officially, the Church called for consultation, for example, a Christian who had been Jewish and who was spreading Judaism for Christianity's sake, but nothing compares to what is narrated in this fiction. Gonzaga (1993), for example, in his work *The Inquisition and Its World,* contextualizes the Inquisition and does not show it as a phenomenon isolated to the values and culture of his time, regardless of what is said about it, just as Voltaire also deals with the subject. The myth of the murderous Catholic Church is a deception that exists among the foolish and uneducated people of our time.

Escarmented, he is taken away by a relative of the Inquisition who cordially lodges him in the Church premises. The inquisitor makes him confess some sin and imputes to him an expense, an obligatory bail for freedom.

This clarifies that the motivation to arrest Escarmentado existed because the representatives of the fictional Inquisition realized that he had some money.

In fiction, they made use of ecclesial authority to profit from the freedom of others, as can be read:

> "At night, when I was about to get into bed, two relatives of the Inquisition arrived with the holy Hermandad; They kissed me tenderly and took me, without saying a word, to a very cool dungeon, furnished with a mat and a beautiful crucifix. I remained there for six weeks,

at the end of which the Reverend Father Inquisitor sent me to ask me to come and speak to him: he confined me for some time in his arms, with a very paternal affection; he told me that he was sincerely distressed to have known that I was so badly housed; but that all the apartments in the house were occupied and I hoped that, next time, I would feel more at ease. Then he asked me cordially if I did not know why I was there. I told the reverend that probably because of my sins. "Well, my dear son, for what sin? He speaks to me with all confidence". As much as I searched, I could not guess: he charitably helped me. Until I remembered my indiscreet words, that I was remitted with disciplines and a fine of thirty thousand reales. They led me to salute the grand inquisitor: he was a polite man, who asked me how I had found his little party."

Towards the nation of Turkey, Escarmentado pondered and proposed not to give an opinion on the festivals of others so as not to suffer from the problem of divergences and intolerance, especially with the intolerance of the Turks who, in his conception, were much more cruel for not having been baptized.

At this moment, Voltaire says that human cruelty does not leave men even after baptism, which characterizes irony, in fact that baptism is, according to the narrator's faith, a rite of entry into the Christian life that does not fit with cruelty. St. Paul once expressed ironic frustration with baptized persons when he said, "Thank God, I have baptized none of you except

Crispus and Gaius." So it is common among scholars to make a timely joke about one's faith.

Scared, he was astonished to realize that Christians lived in the same land as the Turks and for using artifices to keep the peace, such as blaspheming others through a language incomprehensible to the other. Let's see how this presents itself:

> "So I went to them. I was very surprised to see that in Turkey there were more Christian churches than in Candia. I even saw numerous groups of monks, whom they allowed to pray freely to the Virgin Mary and curse Mohammed, these in Greek, others in Latin, others in Armenian. "Good people, these Turks!" I exclaimed.

From the number of Catholic churches in Turkey, it seems that Christians enjoyed conflict, which is ironic. Voltaire criticized and laughed at everything, not only Christianity as some believe, so Voltaire revolted with the Church is another character in the fictions of some essayists and historians.

Voltaire recognized the savagery in peoples that existed without depending on their convictions and acted, in part of his writings, as he always did in his current and real life, to generate playfulness, to satirize and disassociate individuals from their social representations when necessary.

Let us look at Escarmentado's impressions of the savagery of the peoples of the East:

> "To console myself, I took into my service a beautiful Circassian, who was

> the most affectionate of creatures in intimacy, and the most devout in the mosque. One night; in the sweet transports of his love, he exclaimed, kissing me: Alla, Illa, Alla; they are the sacramental words of the Turks: I thought they were those of love (groans of pleasure); I also exclaimed with all tenderness: Alla, Illa, Alla. "Ah! praise be to the God of mercy," she said to me. – Now you are Turkish". I told him that I blessed him for having given me the strength of a Turk, and I thought myself very happy. In the morning the imam came to circumcise me; and, as I was reluctant, the cadi of the neighborhood, a loyal man, proposed that they impaled me: I saved my foreskin and my ass with a thousand sequins, and fled without delay to Persia, resolved no longer to hear Greek or Latin mass in Turkey, and never again to cry out: Alla, Illa, Alla in an amorous encounter."

Let us note that foolishness is, by Voltaire, attributed to all peoples who manifest their own self depending on culture. In Persia, for example, there was a disagreement that was apparently not religious between two sects. There was a rivalry between the white ram group and the black ram group:

> "When I arrived at Isfahan, they asked me if I was for the black ram or for the white ram. I replied that it was indifferent to me, as long as the lamb was tender. It should be noted that the White Ram and Black Ram factions still divided the

> Persians. They thought I was mocking both parties, so that, already at the gates of the city, I found myself involved in a violent quarrel: it cost me countless sequins to get rid of the sheep."

In this case, where there is no apparent religious aspect involved, there is partisan passion, and this is one of the aspects targeted by Voltaire's criticism.

To say that men conflict with others because of their particularities and to contradict these particularities by destabilizing them, is what Voltaire did.

The words of an African captain introduced near the end of the tale reveal one of the main reasons for the tale, which talks about tendencies in favor of conflicting over particularities and seeking adversities that the ignorant have. Let's see what the captain says:

> "You have long noses," replied the black captain, "and ours is boring; his hair is straight, ours is curly; you have gray skin, and we have ebony knots; we must, therefore, by the sacred laws of nature, always be enemies."

When idealism says that one should not conflict over particularities, Voltaire puts contrary words in the captain's mouth. Obviously, this war does not exist naturally if there are no idiots, in the Greek sense, who foment it, which characterizes irony that consists in bringing a different interpretation of reality.

The tale ends with Escarmentado realizing that the best thing to do is to escape the adversities of the world, return to his land and get married to live a quiet life, the same message as his novel *Candide or Optimism*. He returns to the starting point to close the cycle.

After getting married, Escarmentado receives a pair of ornaments and considers that that state where he is betrayed by his wife is the most peaceful he can achieve in this life, creating an aversion to what happened to his father at the beginning of the tale, as an ironic reversal in the story.

The name Escarmentado suggests that our protagonist is someone innocent who suffers without deserving suffering, someone who learns from experience and experience has shown him, by tasting the stupid adversities of the world, that the best thing is to be a quiet betrayed who stays away from adversity.

Thus, Escarmentado, realizing that it would not make any difference in his world, regardless of the attitudes he took, and that the only good he could achieve with conflicts or escape from his reality would be the misfortune of his own life, he took an attitude opposite to that of his supposed father who did not conform and fell into disgrace.

Escarmentado's attitude ironically works against what is conceived as what is certain to happen and what happens in the world and this, having an attitude different from the common ones, brought him good.

The message of Escarmentado is to do things differently. This difference is not in the conflicts of party ideals and that is in taking care of one's own life by accepting the afflictions common to oneself without seeking new ones, by fleeing from the spread of evil or the reaction in a stupid way against it, that is, the difference is in taking care of one's own garden, the same reason expressed in the great novel *Candide or the*

Optimism against the will of most of those who consider themselves wise advisors capable of to make the lives of others illustrious by persuasion.

V - CONCLUSIONS

By this diachronic and synchronic study, I am not the one who will determine idealism about any phenomenon, because I am studying and I must, for honesty, bring possibilities. If I make use of current language, I would say that ironically the semantic charge of 'irony' has undergone semantic inversion in more contemporary studies, so that today, the causes and consequences common to irony are taken for it and not for indications of its existence in cases of clear dissimulations such as the example where the people filmed are warned to laugh, where there is no indication of the phenomenon studied but dissimulation, which makes me dare to say that there is no irony in these cases, however, I consider, obviously, that each one has the freedom and arbitrariness to define the phenomena in any medium as he sees fit.

I set out to bring an opinion contrary to the opinion of the person who evidenced irony to preach it as an *immediate resource of the barbarians; a favorite gesture of the ignorant and the first impulse of stupidity (R. F.)*, after making it clear that the appeal manifests itself depending on the individual character and the momentary intention in disputes. In this way, Socrates tends to develop the phenomenon differently from that of Dionysodorus, because he has a different intention.

As for **literary irony** divided into **verbal** and **situational**, I fully agree with its existence and evidence, because, in cases, irony exists in some way, even if it is not evident as a resource between two or more interlocutors. The same occurs with dramatic **irony** , which is perceptible through elements that precede it, evidencing that it can exist in some way even if not completely verbalized, in a presupposed way when it is part of a planned chain and as a necessary entailed element to exist.

There are characters to be played who mature in a very similar way in the three works analyzed by Voltaire in different ways and obviously not at the same level of naivety and not about the same phenomena, which is clearly inspired by the reality of the author Voltaire. With this, we can speculate that Voltaire wanted to say that he found virtue after making a lot of mistakes, which generated his character Escarmentado as a charity manifested of himself with the purpose of educating and avoiding deception by using his own example, which he learned. This is a clear sign of love and benevolence on your part.

St. Augustine said that writings of very learned men narrate that even shoemakers and men of lower class social conditions devoted themselves to philosophy, who shine with so much light of their intelligence and virtue that they would in no way have wanted to exchange, even if they could, their social position for any title of nobility and Candide, who did not have the nobility recognized, began to recognize the value of a life dedicated to work when he said: "- This good old man seems to have conquered a destiny preferable to that of the kings with whom we had the honor of having supper.". The character Martinho dared to say: "- Let's work without reasoning. [...] It is the only way to make life bearable."

The message is clear: "Let us forget the ramblings and analyses and live in truth, in good work."

In the case of the novel *Candide or Optimism*, the Socratic ancient Greek questioning that maintained all the **verbal**, **situational** and **dramatic** ironies about the phenomenon of optimism existed in the text, that is, the rhetorical resource irony entailed all literary ironies. In other cases, the real referent of the phenomenon may not exist graphically in the text and still exist in some way, since it would be possible to detect its existence by the data presented, that is, it has an existence presupposed by elements that are its consequences.

I prefer to say that the detectable elements that promote irony are ironic, just as I call the consequences of irony ironic, because I consider that irony happens at some time or place, verbalized or not in some way, however I prefer to call irony the Greek phenomenon of the Ancient Age as a rhetorical resource because in it there are at least two narrated consciousnesses where one questions to lead to contradiction or entailed answer, which presents the completeness of the referent of the phenomenon.

In essence, the rhetorical resource of irony is similar in complexity to the questions of the Greek Socratic tradition that sought to expose wrong conclusions before the possible correct ones or the correct conclusion, as it is possible to see in all those well-educated in the tradition, such as St. Augustine of Hippo, who uses questions to say that God is ineffable, indefinable by words, as in the following example:

> What are you, then, my God? What art thou, I ask, but the Lord my God? "Who then is sir, but the Lord? or who is God but our God?" O Most High, infinitely good, most powerful, rather all-powerful, most merciful, most just, most hidden, most present, most beautiful and most strong, stable and incomprehensible, immutable who changes everything,

> never new and never old, innovating everything, leading the proud to decrepitude, without being aware of it, always in action and always at rest, gathering and needing nothing; loading, filling and protecting; creating, nurturing, and concluding; seeking, even though you lack nothing. (Patristics - Vol. 10 - *Confessions - St. Augustine*, Chapter I, 4. *God is ineffable*)

When St. Augustine, Catholic bishop of Hippo, asks: 'Who then is, Lord, but Lord?' he does not want to know where else the Lord is, because he preaches it, so he knew who God is, but what are definitions if not the fruit of human arbitrariness associated with linguistic signs?

The intention of St. Augustine of Hippo is not to persuade God to give him an answer, nor to dissimulate for God to say who he is, but rather, by his own rhetoric, to say that God is ineffable, indefinable, so that the truth is in him and not in words, and it is on this basis that the irony that seeks to evidence the truth manifested itself.

In a speech or monograph, the questions that lead to wrong answers so that, in the end, lead to the correct answer simulate two or more consciousnesses in dispute for the truth.

Despite being one of the best in rhetoric, St. Augustine did not believe that words could illuminate as if they formed the light of truth, which is why God is predicated as ineffable.

This is the distinction between the rhetorical resource, irony, and other rhetorical questions. Obviously, the rhetorical resource of irony would not

The Rhetorical Resource Irony

work with God because God is not persuadable because there is no real need to change His thinking.

The following ironic expression can be used for God when the person referred to is not holy as a playful resource and not for the sake of convincing: 'Oh God, I do not deserve to be close to this servant and holy person, because my presence hurts his actions and untruths. Be merciful to this person and keep me away from him.'

In this example, there is no attempt to convince the interlocutor God, the expression is the result of a tortured soul that perceives a bad and lying person to preach himself as a saint and the request for removal is real. In the case of the person who pretends to be a saint to hear the request, there is an invitation to reflect on one's own attitudes.

The unspoken question exists in some form and it can be: "God, when will you give me the grace to realize that I am free from this stump?"

God has good humor and appreciates it in us. Let's look at these two ironic questions:

> Who among you will accuse me of sin? (Jn 8:46)
> "Can good things come from Nazareth?" (Jn 1:46)

In the first question, Jesus does not want to know the answer, he makes it clear that they cannot accuse him of sin. In the second ironic question, Nathanael perceives the nature of Jesus contrasting with that of the inhabitants of the capital of Israel, so he is sarcastic, there is real offense present and Jesus praises him for his sincerity. That is, in one moment Jesus develops Socratic irony and in a second moment he praises that he developed it, which elevates the character in accordance with that of Christ. Deus vult!

I purposely made a less in-depth analysis of the two stories proposed to show that, when there is the intention to contradict in a joking way, there is also the need to ironize that tends to form literary ironies.

It is certainly possible to say that certain texts are ironic in some way, however, I avoid calling any literary manifestation that has the motive of contradicting irony.

Irony is attributed, nowadays, to many artifices that contradict, so it is up to scholars to restrict their evidence to the ancient Socratic Greek manifestation, to questions that lead the interlocutor to reflect on his convictions, to understand that, if the questioning is not verbalized and exists in some way, the irony phenomenon can exist in some way.

From this perspective, the artifice that aims to make the reader contradict his own opinion would be called ironic along with the semantic and situational inadequacies, as they are necessary elements for the promotion of the rhetorical resource irony.

Taking any partial or complete semantic modification for irony can lead those who critically analyze a work to take authors' mistakes as genius. This mistake can happen more in works that have the proposal of being jocular, because in them the use of irony is predictable and irony predicts the use of a change of semantic load. Knowing this, it is evident that it is important to know the intention of the writer or sender.

An individual who has used his wrong conceptions or those of orality and modified, through them, the semantic load of a phenomenon in a narrative,

The Rhetorical Resource Irony

cannot be taken as a genius but as an idiot in the Greek sense, an individual who uses himself, that is, his own convictions to work on concepts about phenomena.

For this reason, 'idios' from the Greek, which generated our adjective 'idiot', referred to someone outside public life, because it is outside of conviviality that mistakes, private definitions develop, because there is no one to correct and because there are no practical samples, which leads to creating one's own.

Would a few hundred years be equivalent to this kind of isolation for humanity, or are we establishing ourselves as less discerning to the semantic loads of words?

Voltaire's ironies certainly enrich the sample of irony present in literary art, because in them there is a lot of complexity and richness that are the result of education along the lines of Catholic universities that rescued the ancient Greek education, acquired at the Louis Le Grand school by him.

His education endowed him with good dialectics and the ability to create different levels of complexity in writing that provided the development of ancient Greek irony as is noticeable in his work, however, having good rhetoric is not an indication of knowledge and the type of irony worked by Voltaire in his great novel does not aim to inform, It aims to entertain and make people laugh about the main phenomenon and, nevertheless, there are personal messages between the lines as seen.

The ludic phenomenon or *game* is present in literature on the basis formed by its own time and place that is neither current life nor real life, however, despite signing a pact of verisimilitude with limited time to be undone, we can assimilate fictional values, be led to reflect on our own convictions and, in the relief provided by the *game*, we have what is necessary to complete an idea.

One does not read *Candide or Optimism* to verify the truth of life, but to laugh at it, although with the proper education one can get much more than laughter from such a rich work.

Plato also laughed at the Sophists and the world and made his readers aware of great evils. As was verified, at the beginning of this one, with the brothers Euthydemus and Dionysodorus who did not offer a real sample of knowledge, used trickery, fell into contradiction and were declared winners in the dispute against Socrates and his and in this there is a certain irony in the situation and promotion of the idea that the majority is cheaters and favors cheaters, for having been the majority to declare the brothers as winners.

Voltaire used semantic inversion when defining 'optimum' (best possible) among other definitions, that is, he used semantic modification to ironize, which makes it evident that, in his great novel, he did the same as he did in his practical life, in his quest to satirize definitions formed about phenomena, representatives and ordinary people.

In the novel studied, Voltaire laughed at the real contradictions that are possible in life, so I think that one of Voltaire's great messages in *Candide or Optimism* and in his ironic works is that one should not fully believe in a reason that is at the mercy of man's will, one should not attach titles to patterns of behavior or competences, because vices often override the reason of man, who is a flawed being and who often seeks to find a way to explain errors or vice itself with his characterization or by using rhetoric similar to that of

The Rhetorical Resource Irony

the sophists, that is, man should not be trusted. A noble message, hard to find and much harder to apply wisely.

It is notable that the characters of the nobility, presented in his great novel, do not have names, but noble titles equipped with the characters of their owners who are not noble, which corroborates what has been said about their tendency to disassociate titles and patterns of behavior. Thus, Voltaire understood that what you really are is what defines you and not inherited and purchased titles.

Many laymen will find Voltaire rebellious or adapted to what they want him to be. I found the true one defined by himself.

Possibly it is the message of the most mature Voltaire's innermost self, the message of the scorned man who returns to his home and accepts the afflictions of the world as if they represented the best possible; of the one who returns, at the end of life, to the sweet lap of the Father; of Cândido who abandons digressions and analyses and dedicates himself to his work without thinking; of the mature one who wants to take care of his own garden.

It is common for the productions of a writer to advocate in favor of him and, in this conclusion, there is evidence of Voltaire's self-criticism, as he was, in part, justifying himself.

It is known that in-depth studies on rhetorical resources are necessary for a greater understanding of complex narratives, and the rhetorical resource irony is the least understood resource, especially when it does not contextualize the phenomenon studied.

This dissertation contributes to the notion that it is necessary to study the phenomena involved in advance, that is, it is necessary to develop a study plan that involves the phenomena mentioned in order to have the evidence of the truth, because, if this is not the case, due to ignorance, Voltaire is taken for a revolted quarrel for any reason or a Voltaire adapted to himself who serves to interpret his own desires.

As an example of this, most without a complete reading of themselves have a particular Voltaire. There is the anti-Catholic, agnostic, Freemason, Rosicrucian, Liberalist, Deist and many others Voltaire depending on personal foolishness, except for Voltaire in an assertive timeline with reality itself.

I am aware that there is a great complexity in the phenomenon present in the works studied that is difficult to understand for most, which creates the need to go to the correct sources to have an almost complete notion of the use of the resource.

If it is about a phenomenon, one must study its real samples in a diachronic and synchronic way; if it is about *Theodicy Essays*, one must read *Theodicy Essays* and if it is contextualized in a certain time, one must observe that time.

It is not possible to adapt a definition to its referent without offering a contextualized sample, even so, among the ignorant, there are those who conceive the foolish idea that a complex phenomenon such as irony can be understood with a few lines of a dictionary.

They conceive the acquisition of knowledge in this way because they are not wise, because they do not understand how the world works and, consequently, without experience, they do not understand textual interpretation and the more complex phenomenon of literary art that may require, to be understood, in use, many theoretical and historical bases, in addition to real samples that promote the adequacy of the definition with its real referent so that, finally, in experience, we can recognize the phenomenon more fully.

In an expression where there is only a noun linking verb and its predicate, there are only words to predicate others. So, how to understand the referent only by words?

Let us take the example of St. Augustine who, despite having good rhetoric and defining God well with words, was not overwhelmed, because he defined the real referent of his words as ineffable, indefinable.

The dedication to the study of rhetoric brought, in these contextualized interpretative analyses, by following, in part, the Aristotelian method of defining, an addition to the understanding of the phenomenon under study in the literary art of Plato that narrates the wisdom of Socrates and in the literary art of Voltaire, allowing us to have a small notion of the wisdom of the great philosophers and the genius of the great representative author of French literature whose literary production contributes greatly, for its richness, for studies on rhetoric and, consequently, on the irony that was the analytical category used.

There is a need to study the resource to avoid deception, because it is possible to use it to disseminate, in popular knowledge, wrong ideas.

In the works studied, the rhetorical resource of irony allowed the creation of several semantic inversions and distinctions between nature and what is said about something, provoking laughter or ridicule through the semantic and situational inadequacies necessary for the resource to exist.

In the novel *Candide or Optimism*, for example, the reader is led by the narrator to have the same conceptions as him in a space that does not belong to ordinary and real life, as Johannes Huizinga explains in his work *Homo Ludens* about *the game* in literature as an escape from reality in favor of a relief from the tensions of current and real life.

In two of the works cited, Voltaire's narrative voice cites family members of the inquisition who clearly work to monitor the lives of others. In the novel, a family member takes Pangloss for speaking his mind and Candide for listening approvingly and, in the short story, a family member takes Escarmentado to be robbed through a bail. Passages like these affect the reality of the less discerning, of psychologically fragile people who need to have something against Catholicism and find in fiction something to support their hate speech when they should be entertaining themselves.

Ignorant people who don't know how to appreciate literary art, they tend to enter an *easy game* where there is a bad side and a good side and, once they choose one side, they define it as good and dump all the guesses on the other side.

Historical and literary samples are important, because by ignoring them, many consult characteristics of libertine philosophers and attribute them to Voltaire because he was a great representative of French literature, which can cause wrong attributions against him as many have already done. The most recent one I found attributed to him more harmful characteristics of a group that did not represent him and called him a thief.

In fiction, there is no discernible distinction between the holiness of the Church and the dirty character of the characters linked to it, because probably, while studying, Voltaire noticed the behavior of traitors of the type of debauched heretics and this certainly inspired him. In relation to laymen, in *Memnon or Human*

Wisdom, for example, it is concluded in fiction that moderate vices are part of life, which can lead unwary and psychologically fragile people to some deception.

Because of the misunderstandings, many untruths arose orally about Voltaire. Most people do not consult reliable sources and are content with empty slanders that would not hold up to the truth.

Knowledge of dictionaries may not provide a true study of the proposed phenomena, so without a true study, it is difficult for the naïve to distinguish truths from fallacies present in orality and writing.

Voltaire, with his ironies, contributed many of them to lead, without realizing it, the unprepared and psychologically fragile to conclude erroneous ideas and I think that all remarkable people, good or bad, lead psychologically fragile people to abysses. Voltaire would not be a source of good for all people who read him. Today they interpret Jesus for villainous practices, why wouldn't they do the same with Voltaire?

The mistake is such that, during my time in a university environment, I heard some professors repeat elements that only exist in Voltaire's fiction as if they were historical truths and imagined elements about the author.

Let us bear in mind the possibility that there is an intention to deceive when we hear or read criticism, because many of them are well elaborated to entertain or spread some wrong idea and not to inform the truth. Let us also bear this intention in mind when we hear arguments, for inadequacies are designed to deceive like the inadequacies of the sophists who won the contest against Socrates and his followers in *Euthydemus*.

It worries me to see many unwary people take certain ironies of Voltaire's fictions as truths, which makes me think that this is due to the lack of educational basis. One must read *Theodicy's Essays* to understand that Voltaire was ridiculing a straw man of Leibniz's. The referent is in the work and not in what is said about it, and the one who talks about it must prove what it says when it dedicates itself to quoting.

Foolish people cannot admit the truth in the observance of their own reasons, for they place wills above all and everything and their vices as the grace of a god. This leads to confusion where it is not possible to distinguish truth and vice. Someone who takes the truth as he pleases cannot recognize high values because of his materialism and there are many who talk about what they do not know for the sake of creating a revolutionary liberating Voltaire because they condone joking elements that were not even his real ideas and that served to entertain and not to transmit truth.

Narrating a homosexual priest and swindler does not imply that Voltaire thought that all priests were homosexuals and swindlers, but that the phenomenon exists in some form and that the character of the ordained traitors, in the same way the cowardly philosophers narrated, is jocular, since it is evident that he had respectable friends in philosophy and faith, as is observable in your life.

The elements that aim to evidence the truth are clear, the irony used by Socrates to evidence the truth always makes it possible to adapt words to truth, that is, to adapt linguistic signs with their semantic loads to their referents in a contextualized way.

Because of his wisdom, obviously when necessary, Socrates, whose character I could define as one of the highest personifications of the human race, should be used as a model for irony.

The Rhetorical Resource Irony

Is this the best of all possible worlds? Can good things come from Nazareth? Those closest to Euthydemus and Dionysodorus will refute any answer given, they will position themselves on the opposite side regardless of the truth to be right about other consciousnesses. Those closest to Socrates will raise questions in order to remove all the wrong possibilities to find the right one.

Let us note that those closest to Euthydemus and Dionysodorus can be on both sides of opinion on any subject and it is these who have their idealisms destroyed under certain circumstances.

On the one hand, what moves action is to win over beings of the human race, and on the other hand, the motive is to achieve truth, virtue, excellence.

Socrates spoke about a peace of mind acquired by always being honest and therefore developed the rhetorical resource irony in a very specific way alongside the truth.

'σημαντικός' , 'sēmantiká';
Study of signification
on the rhetorical resource irony.

REFERENCES

AUGUSTINE, Saint. *Patristics - Vol. 24 - Against the Academics – The Order – The Greatness of the Soul – The Master.*, Paulus, 2014

AUGUSTINE, Saint. *Patristics - Vol. 10 – Confessions - St. Augustine,* Paulus, 2014

CATECHISM OF THE CATHOLIC CHURCH, 2ND ED. Publisher: Vozes, São Paulo, 1999

CORTÁZAR, Júlio. *Some Aspects of the Tale*, Valise de cronópio, São Paulo, Perspectiva, 1974.

CSISZAR, Sean Anderson. *Luther's Golden Book* – Includes the 93 theses, 2015

GONZAGA, João Bernardino Garcia. *The Inquisition in Your World*, ed. 4, São Paulo, Saraiva, 1993.

HOPS, Daniel, *History of the Church* (10 volumes) Volume III – The Church of the Cathedrals and the Crusades, ed. 1 (from the French Academy)

HOUAISS, Instituto Antonio. Houaiss (dictionary), Editora Objetiva, 2009

HUIZINGA, Johannes, *Homo Ludens*, The Game as a Cultural Element, Ed. Perspectiva, 2014

JOSEPH, Sister Miriam, *The TRIVIUM*: The Liberal Arts of Logic, Grammar, and Rhetoric, 2nd ed., Achievements: 2014

LEIBNIZ, Gottfried Wilhelm. *Theodicy Essays* on the goodness of God, the freedom of man and the origin of evil, trans., intro and notes: William de Siqueira Piauí and Juliana Gecci Silva, Ed. Estação da Liberdade, 2013.

LEITE, Lígia Chiappini Moraes. *The Narrative Focus* (or the controversy around illusion), 10th ed., 5th printing, Ática, São Paulo - SP, 1985

MARIA, Fr. Júlio. *The Devil, Luther and Protestantism*, 2nd ed. – Mahnhumirim - Minas – 1950.

MAUROIS, André. *O Pensamento Vivo de Voltaire*, São Paulo: Martins, trans. Lívio Teixeira, 1975

MICHAELIS: Modern dictionary of the Portuguese language. São Paulo: Companhia Melhoramentos, 1998

MUECKE, D. C. *The irony and the ironic*. Trans. Geraldo Gerson de Souza. São Paulo: Perspectiva, 1995

PLATO. *Euthydemus*, Text established and annotated by Jhon Burnet; Translation, presentation and notes by Maura Iglésias. Rio de Janeiro; Ed. PUC Rio; Loyola, 2011.

VOLTAIRE. *Candide or Optimism*, Trans. Roberto Gomes, Ed. E&PM Poket, vol.92, 1998

VOLTAIRE. *Breves Contos II,* 1st ed., LL Librari, 2013

(Photo by the author in front of the Cathedral Basilica of Our Lady of the Snows,
João Pessoa – PB, 07/05/2022, 9:18 pm, during priest ordination.)

JONAS BATISTA DOS SANTOS is Catholic and graduated in Letters from UFPB. Born in 1980, in João Pessoa-PB, he dedicates himself to narrative fiction, poetry, philosophical essays and research in linguistics. His work moves between the rigor of grammatical analysis and literary creation, guided by the search for truth, virtue and excellence, pillars of good rhetoric. Inspired by Plato, Aristotle, Socrates and Voltaire and moved by the Catholic perspective of language as an expression of the Logos, he proposes a reflection on the adequacy between words, thought and reality. This book is the result of her path of study, teaching and practice of the Portuguese language. **Publications:**

Rhetoric and Linguistics: *The Rhetorical Resource Irony (2016), Rhetorical Adequacy (2026)*
Philosophy: *Clinyas (2016), Maximo (2018), Solitude (2026)*
Politics and education: The Communist League (2016)
Doctrine and apologetics: VERUS DEUS (2016), THE Heretic (2026)
Moral Psychology: Monstrum (2016)
Eschatology: The Village (2020)
Medievalist Courtly Love: The Rose (2020)

www.ingramcontent.com/pod-product-compliance
Lightning Source LLC
Chambersburg PA
CBHW031627210526
45464CB00004B/1782